D0788992

# Landscaping on the New Frontier

# Landscaping on the New Frontier

## Waterwise Design for the Intermountain West

Susan E. Meyer
Roger K. Kjelgren
Darrel G. Morrison
William A. Varga

Illustrations by Bettina Schultz

UTAH STATE UNIVERSITY PRESS
Logan, Utah
2009

Utah State University Press
Logan, Utah 84322-7800



Manufactured in China

ISBN: 978-0-87421-709-4 (paper)
ISBN: 978-0-87421-710-0 (e-book)

Cover design by Barbara Yale-Read

Photo credits:

*Introduction*: Waterwise home landscape photo—Linda Oswald.
*Design*: Wellsville Mountains photo—Roger Kjelgren; watercolors—Hayley Olsen, Jennifer Hale, Sandy Blackner, Talia Poulson.
*Installation*: Stone path and bridge photo—Phil Allen.
*Native Landscape Pioneers Tell Their Stories*: Photos provided by the authors.
*Plant palette*: New Mexico privet photo—Roger Kjelgren.
*Chapter display*: pp. xiii, 1, 71, 96, 119 —Dan Miller.
All photos not listed above are by Susan Meyer.

Library of Congress Cataloging-in-Publication Data

Meyer, Susan E.
Landscaping on the new frontier : waterwise design for the Intermountain West / Susan E. Meyer, Roger K. Kjelgren, Darrel G. Morrison ; illustrations by Bettina Schultz.
p. cm.
Includes index.
ISBN 978-0-87421-709-4 (cloth) -- ISBN 978-0-87421-710-0 (e-book)
1. Native plant gardening--Great Basin. 2. Natural landscaping--Great Basin. 3. Landscape irrigation--Great Basin. I. Kjelgren, Roger K. II. Morrison, Darrel G. III. Title.
SB439.24.G74.M48 2009
635.90979--dc22

2008051557

*This book is fondly dedicated to*

Dr. Kimball T. Harper

*long-time student of intermountain native plants, ecologist extraordinaire, and father of the intermountain native plant nursery industry.*

*Silver daisy*

# Contents

## Chapter Six
## Native Landscape Pioneers Tell Their Stories 131



## Chapter Seven
## The Plant Palette 159

*Maple mallow*

# Preface

On first glance, a landscaping book that mentions "waterwise" in the title suggests that water conservation would be a major theme. But, while this subject is certain to become more pressing as population growth inevitably intersects with a finite supply, it's not just about water. Traditional cultivated landscapes in the Intermountain West—carpet lawns, isolated trees, rows of shrubs, and decorative patches of flowers—are borrowed. True, such landscapes serve a variety of functions and appeal to a tidy aesthetic. But "Norway" maple, "Austrian" pine, "Chinese" juniper, "Kentucky" bluegrass, and "Persian" lilac all illustrate that something is missing. That something is the Intermountain West.

To anyone who has marveled at the natural beauty evident in our wild landscapes and felt a desire to capture even a small portion of this beauty in a home landscape, *Landscaping on the New Frontier* is written just for you. To the skeptic disappointed by the notion that native plant gardens are doomed to become "a patch of weeds or a pile of rocks," prepare to be pleasantly surprised. And to all who at one time or another have assumed that native landscapes are either completely carefree (i.e., because wild places look beautiful without a gardener) or are essentially like traditional landscaping, just with different plants, prepare to be educated, with the assurance that the journey will be richly rewarding.

Just as communities of people—with diverse skills and contributions—unite to make a better place to live, the concept of plant community is critical in designing intermountain landscapes. Because our region is characterized by a wide diversity of landscapes, a basic knowledge of these plant communities is essential in designing a yard that will express your creativity while capturing the beauty of native landscapes and respecting the real-world conditions of your property. Thus, much of the thrust of this book is aimed at helping you to understand the native plant communities that are available for inspiration. Subsequent chapters provide the practical how-to's of design, installation, irrigation, and maintenance. A plant palette section provides details on native species that have been shown to thrive when used appropriately in home landscapes in the Intermountain West. And for those who need a little extra inspiration before taking the plunge, seven genuine, intermountain native, home landscapes are highlighted by the gardeners who created them.

Gardening on the new frontier may initially seem strange and even radical to those accustomed to the sameness of most traditional landscapes—"What do you mean I shouldn't add topsoil before planting desert shrubs?" But the complexity, beauty, and flexibility of an intermountain native landscape make it immensely satisfying, and native gardening can be so much more interesting than traditional gardening. If you didn't at least partially believe this, you probably wouldn't be consulting this book (except, of course, in the interest of reducing the amount of water required in your yard). But, as even a brief perusal of the consistently enthusiastic testimonials of those who agreed to share their stories will illustrate, venturing into the new frontier does much more than conserve water. Indeed, even modest success in this effort will meaningfully connect the living space immediately surrounding your home with the beauty inherent in native plant communities. And, for me at least, the ability to leave my yard for days or weeks in the summer and not worry about what it will look like when I return is definitely a bonus.

Dr. Phil Allen
Department of Plant and Wildlife Science
Brigham Young University

*Western columbine*

# Acknowledgments

In the journey that led to this book, we had many guides and helpers along the way. Our book was conceived as a companion volume to *Water Wise: Native Plants for Intermountain Landscapes*, published by Utah State University Press in 2003, and we owe a debt of gratitude to the authors of that book, especially Wendy Mee and Jared Barnes, for leading the way toward an all-native intermountain landscape aesthetic.

Jennie Bear and Dale Torgerson were involved in early stages of development of the design graphics for this book.

We want to thank the following individuals for "beta-testing" the penultimate draft of the manuscript: Cindy Turner, Alisa Ramakrishnan, Stephanie Carlson, and Tara Forbis. We also thank Phil Allen and Larry Rupp for providing helpful professional reviews, along with two anonymous peer reviewers.

The section of the book entitled "Native Landscape Pioneers Tell Their Stories" was authored by several different people who brought a variety of perspectives to their essays. We would like to thank them for responding to our request for participation in spite of a short turnaround time and for doing such a good job of describing their experiences: Phil Allen, Jan Nachlinger, Randall Nish, Carl Dede and Angie Evenden, Ann DeBolt and Roger Rosentreter, Roger Kjelgren, and Bettina Schultz.

We are grateful to the following organizations for responding to our call for grant support to help defray printing costs for this book, enabling us to keep the price within reach of ordinary people: Utah State University Center for Water Efficient Landscaping; Utah Native Plant Society Utah Valley Chapter; Intermountain Native Plant Growers Association; Central Utah Water Conservancy District; Jordan Valley Water Conservancy District; and Weber Basin Water Conservancy District.

We especially want to thank Michael Spooner of Utah State University Press for inviting us to write this companion volume, and for his patient shepherding of the project throughout its long evolution.

Lastly, we thank Bettina Schultz, who not only did an excellent job illustrating the book, but also provided countless hours of informal editorial service, as well as emotional support and counseling along the way.

Prince's plume

# Introduction

**B**ig skies filled with dramatic clouds and magnificent sunsets....vast expanses of dry desert, monumental mountains, fantastic red rock formations and canyonlands that fade into a purple and blue infinity.... meandering streams lined with red osier dogwoods, willows and cottonwoods that turn brilliant red and yellow in the fall.... dry mountainsides blanketed with golden grasses, silver-green rabbitbrush and sage, dark dots of pinyon pine and juniper drifting down their steep faces; dense, dark spires of white fir covering north-facing slopes and cascading down drainage ways.... groves of gambel oaks with twisted trunks and branches blanketing lower slopes; October mountainsides seemingly on fire with the flame-like foliage of masses of bigtooth maples that trail down ravines; ancient groves of cinnamon-barked mountain mahogany; multitudes of quaking aspen with luminous golden leaves fluttering in the fall breezes; undulating mountain meadows richly patterned with a tapestry of grasses and sedges with sprinkles of the jewel-like colors of wildflowers drifting through them.

All of these are memorable images of the Intermountain West. It is a rich and varied landscape, one that attracts visitors from all over the world and has attracted a parade of settlers over the past century and a half who came to live, work, and play in this magnificent landscape. Ironically, in the areas where these settlers "settled"—where they built their homes, schools, churches, business places, and even their city parks—they replaced the richly diverse, distinctive native landscape with one that has its aesthetic roots in England or the eastern United States. This "new" landscape has been one of green lawns, clipped hedges and shrubs, imported shade trees, and beds of dark green "ground cover."

The demise of the native landscapes in the urbanizing Intermountain West is an understandable phenomenon. Nostalgia plays a role; we take comfort in bringing something from our past when we settle in a new and strange place. This familiarity also means we know how to manage the imported landscape, even if that management requires substantial inputs such as fertilizer, fossil fuel, and water in order to make the landscape acceptable. People do love the

A waterwise landscape with many native plants (top) contrasts strongly with a traditional landscape (bottom).

wild, native landscape as a place of retreat and even solace, but they've never been shown how to experience it "up close and comfortable" in their daily lives. They have somehow been led to believe that they should control nature and the native landscape, that they should "improve" on nature. This feeling is stimulated by social pressure ("Whose lawn is the greenest, the most weed free?") and is reinforced by powerful advertising that promotes a standardized, generic landscape all across the country.

This standardization brings with it a number of costs. There are, first, the direct, out-of-pocket economic costs: the resources necessary to create and maintain a lawn, for example, feeding and irrigating the turf to make it grow in an environment very different from the moist environments of England or the northeastern states, then expending energy to regularly cut that turf to a short, uniform, "neat" height, once it has grown. There are environmental costs as well: drawing on increasingly limited water supplies that might better be used for drinking and for food production, using supplemental nutrients that can later become pollutants to support lawns and introduced ornamental plant species, using fossil fuels to mow vast areas of lawns, to clip hedges, and even to blow fallen leaves from one place to another.

But perhaps the biggest costs lie in the ultimate impoverishment of the landscape: the systematic replacement of a biologically rich tapestry of adapted native vegetation with a very simplified and species-poor landscape of plants from other places. Along with the diminution of plant species that occurs in this process, there is a parallel loss of richness from other living things—songbirds, and butterflies and bees, for example. And from our standpoint as living human beings, there is an experiential, aesthetic impoverishment that accompanies this ecological impoverishment. If the designed-and-managed landscape looks essentially the same wherever we are, we lose our sense of place, of knowing where we are. When we homogenize the human-dominated landscape everywhere, "there is no 'there' there," to use the words of Gertrude Stein.

The winds of change are blowing, however, here in the Intermountain West. This is stimulated in part by a growing recognition of the need to conserve water. Existing or potential water shortages associated with an expanding human population, combined with the possibility of increasingly frequent years of low rainfall,

Mosaic of aspen and evergreen forest communities at Kolob Overlook, southern Utah.

raise serious questions about the flagrant use of water to support an English-inspired landscape (average annual precipitation of thirty to forty inches) in a region with an average annual precipitation of only ten to fifteen inches.

The winds of change are blowing too, with people's increasing awareness that there are positive alternatives to long-practiced landscape habits. To further that awareness, it is our intent in this book to make it very clear that it is possible to design and manage landscapes that are experientially rich for their occupants, ecologically and environmentally sound, and aesthetically pleasing. We want to show that it is possible to distill the aesthetic qualities of the richly diverse native landscapes of the Intermountain West and incorporate them into our designed landscapes, landscapes that will restore the sense of place that has been too widely eradicated.

In this book, you will learn how to use natural landscapes to inspire your designed landscapes—how to translate the spatial structure and plant community patterns of the natural landscapes of the Intermountain West into landscape designs that celebrate its difference. We'll show you examples of what some people have already done to convert traditionally-designed landscapes into alternatives that are more expressive of the place, and at the same time more experientially rewarding. And we will discuss an approach to irrigation that

The wide open spaces at the Desert Experimental Range, western Utah.

minimizes the use of supplemental water, yet ensures the survival of plants during unusually dry periods. Thus you will learn how to combine ecological principles with design principles to create beautiful, low-maintenance landscapes that require only minimal resources.

The West, including the Intermountain West, is a region where rugged individualism has been valued, from early explorers like Lewis and Clark, who traveled here two centuries ago, to the pioneers who came westward in the nineteenth and twentieth centuries, to those who are still arriving in the early years of the twenty-first century. Perhaps it is time for a new kind of pioneering to take place in our approach to the landscapes we create for ourselves. Rather than continuing to mimic or imitate the English landscape in our settlement, and spend substantial resources for maintenance, maybe it's time to celebrate the unique attributes of the West, to advance a landscape aesthetic that is of the place, in both its ruggedness and its subtleties. The starting point, the inspiration, lies in the natural landscape itself. It lies in the rich array of plants, the patterns and the process: on the mountainsides, in the deserts, salt flats and canyons, in the valleys and along the meandering streams.

*Leo penstemon*

# Native Landscapes of the Intermountain West

Chapter One

To design beautiful and functional native landscapes, the first step is to learn to look at landscapes in nature and to begin to understand why they look the way they do. Even intuitively obvious truths about intermountain landscapes need to be given some thought. For example, all westerners know that, to escape the heat of summer, a picnic in the mountains is generally a good approach. In the winter, we know that we can head for the desert to escape from the snow. Plants respond to these climate differences at least as much as people do. The native plant communities in high mountain valleys are completely different from the plant communities in the desert country, where people often go to seek winter sunshine.

As you drive up into the mountains from towns nestled in the valleys at their feet, first the low sagebrush steppe vegetation gives way to foothill communities characterized by small trees like gambel oak and bigtooth maple, or to a pygmy evergreen forest made up of juniper and pinyon pine. Further up, patches of quaking aspen and white fir or lodgepole pine start to appear, interspersed with meadow communities of grasses, low shrubs, and an abundance of wildflowers. If you are driving up a canyon with a year-round stream, you will see the difference right away between the streamside vegetation, which is very green and lush, and the hillsides above, which support shrubs and grasses found in much drier environments. Often, arriving in the aspen/white-fir or lodgepole pine zone is enough to relieve the heat of summer, but if the road continues to wind upward, it will pass through evergreen forests of sub-alpine species of spruce and fir, until at last it reaches timberline and breaks out into alpine tundra, the dwarf community that lives on the high, windswept ridges that are too harsh to support trees.

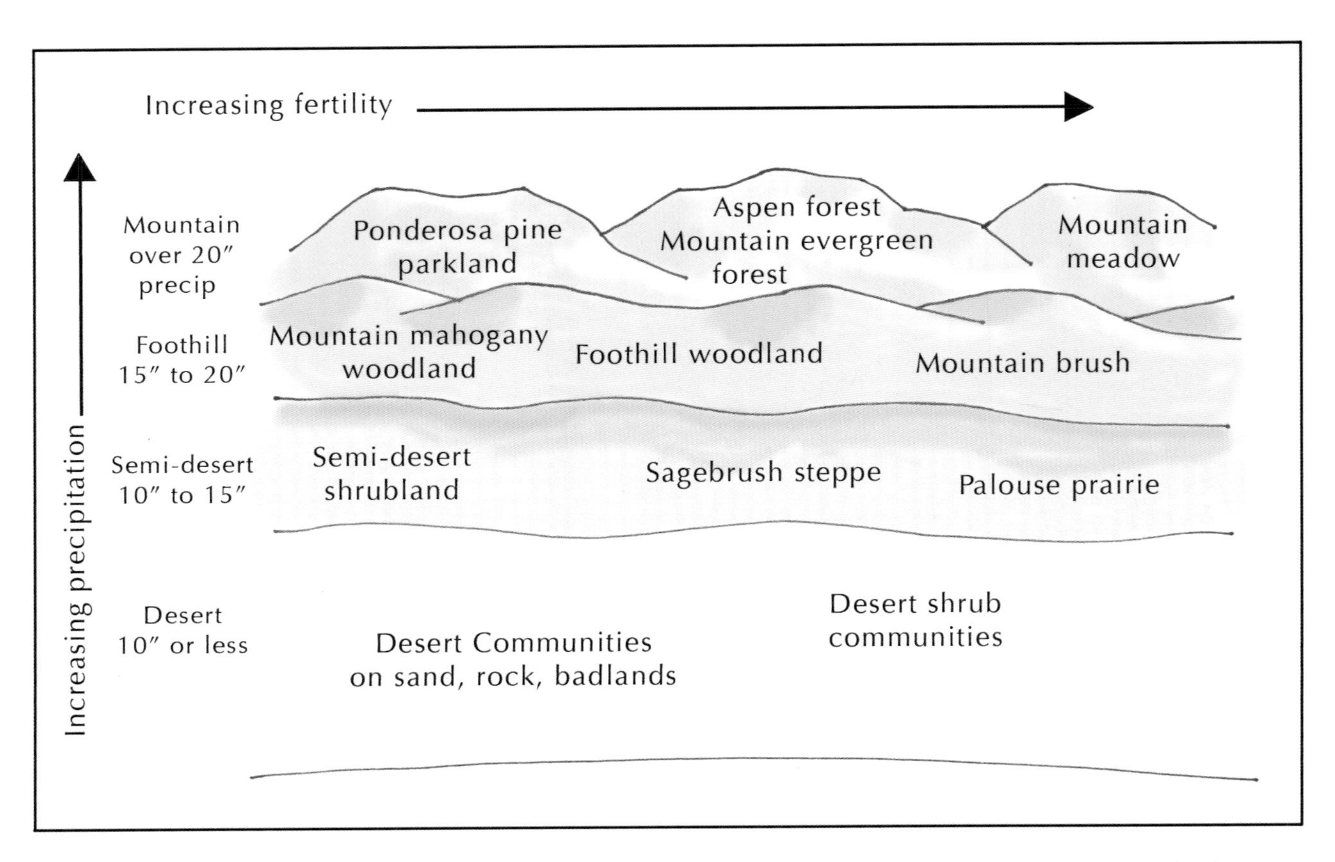


Schematic diagram showing how plant communities in the Intermountain West are distributed relative to the gradient of increasing precipitation associated with increase in elevation, shown on the vertical axis, and relative to the fertility gradient associated with soils, shown on the horizontal axis. Precipitation ranges for the four water zones are shown on the left.

Each of these plant communities represents a response to a set of environmental conditions that define the habitat for the species that occur in that community. By understanding how plants interact with environmental conditions in nature, you will begin to see how you might create designed landscapes that capture the essence of these natural landscapes. You can use suites of species with complementary needs as well as complementary aesthetic features, and group them into patterns that reflect the natural patterns you have observed.

A hallmark of the Intermountain West is its great variability in terms of climate, topography, and geology. The basic theme is that of a generally semi-arid region with cold winters and dry summers, but there are a multitude of variations upon this theme. Majestic mountain ranges rise up like islands out of the Great Basin desert lowlands, while mighty rivers dissect immense canyons into the giant staircase of high mesas on the Colorado Plateau, and the massive spine of the Sierra-Cascade axis creates rain shadow effects far to the east. All this topographic diversity creates an incredible array of growing conditions for plants, and a corresponding diversity in plant communities.

As mentioned above, the most basic climatic trend in the region is related to elevation. Dry environments with hot summers in the desert valleys give way to successively cooler and moister conditions as you travel up into the mountains. Superimposed on this basic pattern are microclimate variations created by differences in slope and exposure. A close look at a west-facing range like the Wasatch Mountains will reveal the great importance of exposure in our region. At higher elevations, the north-facing exposure can be cloaked in white fir and aspen, while the south-facing exposure supports a mountain brush community usually found on warmer, drier sites.

White fir forest cloaks the north slope in a side canyon at Zion National Park, while the south slope supports a drought-hardy mountain brush community.

Look lower down on the mountain, and you will see that the mountain brush community occupies the north-facing slopes, while the south-facing slopes support the characteristic low shrubs and bunchgrasses of the sagebrush steppe community. Each plant community occurs under a characteristic climatic regime, but that regime is the result of the interplay of a number of factors, such as latitude, elevation, exposure, and slope.

Light is another factor that has a major impact on plant communities. In fact, the differences caused by exposure are largely due to differences in the duration and intensity of sunlight. Northern exposures are shaded for longer periods than southern exposures, especially during spring and fall, at least in the Northern Hemisphere. Topographic relief can have an even more dramatic effect on light in places where vertical cliffs and deep canyons are part of the landscape. In a canyon bottom, the ground surface may be shaded for most of the day, and the plants that grow there, while enjoying moderated temperatures and better moisture, must be able to tolerate low light intensity. Rock outcrops in the desert can have a similar effect. Lastly, the plants themselves can change the light environment for associated species. For example, the shade under the closed canopy of an evergreen mountain forest is so dense that only a few species are shade tolerant enough to grow there, whereas the dappled shade created by an aspen forest supports a very diverse suite of understory species.

Over a region as vast as the Intermountain West, latitude has a major impact on climate, particularly temperature. At a given elevation in the southern part of the region, temperatures will generally be much warmer than those at that same elevation in the northern part. For example, St. George, Utah, and Boise, Idaho, have similar elevations, but St. George is in the Mojave Desert, which has relatively warm winters and very hot summers, while Boise, five hundred

The Intermountain West, showing major geological regions (Columbia Plateau, Great Basin, and Colorado Plateau) and population centers in and around the area.

miles farther north, has much colder winters and cooler summers. The elevation at timberline is also much lower at more northerly latitudes than it is in the south, so plant communities are shifted downward in elevation as one goes farther north. This makes it hard to predict climate in a particular place just from a knowledge of its elevation, but the trend for cooler, moister environments at higher elevations can be seen throughout the region.

Climate information for some cities around the Intermountain West. *Locations are color-coded according to their water zone: yellow-minimal water, peach-low water, green-medium water, lavender-high water.*

| | | Mean Annual Precipitation (inches) | Mean Summer Precip | % Precip in Summer (Jun Jul Aug) | Mean Jan Min Temp (F) | Mean July Max Temp (F) |
|---|---|---|---|---|---|---|
| Spokane | WA | 16.1 | 2.4 | 15% | 22 | 84 |
| Yakima | WA | 8.2 | 1.2 | 15% | 21 | 88 |
| Walla Walla | WA | 16.6 | 2.1 | 13% | 28 | 88 |
| Bend | OR | 11.6 | 1.9 | 16% | 22 | 82 |
| Burns | OR | 10.8 | 1.5 | 14% | 15 | 86 |
| Boise | ID | 11.7 | 1.4 | 12% | 22 | 91 |
| Pocatello | ID | 11.5 | 2.2 | 19% | 15 | 88 |
| Elko | NV | 9.6 | 1.6 | 17% | 11 | 91 |
| Ely | NV | 9.6 | 2.1 | 22% | 10 | 87 |
| Reno | NV | 7.3 | 0.9 | 12% | 21 | 91 |
| Winnemucca | NV | 8.3 | 1.1 | 13% | 17 | 92 |
| Cedar City | UT | 10.6 | 2.6 | 25% | 17 | 90 |
| Logan | UT | 17.7 | 2.7 | 15% | 16 | 87 |
| Moab | UT | 9.0 | 2.1 | 23% | 18 | 98 |
| Salt Lake City | UT | 15.6 | 2.4 | 15% | 20 | 93 |
| Cortez | CO | 13.0 | 3.2 | 25% | 13 | 89 |
| Grand Junction | CO | 8.9 | 2.1 | 24% | 18 | 93 |
| Flagstaff | AZ | 21.2 | 5.9 | 28% | 16 | 82 |
| Page | AZ | 6.5 | 1.4 | 22% | 26 | 97 |
| Albuquerque | NM | 8.8 | 3.5 | 40% | 23 | 92 |
| Farmington | NM | 8.2 | 2.4 | 29% | 17 | 92 |
| Santa Fe | NM | 13.8 | 5.4 | 39% | 18 | 86 |

Cottonwoods mark the channel of Comb Wash in southeastern Utah, while big sagebrush dominates the dry upland areas.

Another important effect of topographic relief is seen in the way that the water that falls as rain and snow in the mountains is redistributed. The fact that this water is channeled into streams and subsurface aquifers, where it can be captured and harvested by people, has made settlement of this generally semiarid region possible. And it is still this mountain water that provides for irrigation of farmland and urban landscapes in the desert valleys below, not to mention supplying potable water for direct human consumption. This natural redistribution of water also has important implications for native plant communities. On steeper slopes, less water is available to plants, because more water will run off before it has a chance to penetrate the soil. The soils also tend to be shallower on steep slopes, reducing water availability even more. For plants growing at the foot of the slopes, namely along the channels that collect water from adjacent slopes, water availability is greatly increased. Stream channels often support riparian (riverside) plant communities, made up of species that have much higher water needs than could be met by rainfall alone in a given location.

Geology also modifies the effect of climate on plants. First, geology on a grand scale determines the landforms that develop in a region—range after range of mountains separated by broad, sediment-filled valleys and salt playas in the Great Basin, vast lava plains punctuated by deep canyons on the Columbia Plateau and Snake River Plains, and giant flat-topped mesas of sandstone, shale, and clay on the Colorado Plateau. These landforms, in turn, affect large-scale weather patterns and create the effects of elevation and topography that we have already noted.

On a geological scale more immediately important to plants, different types of rock weather into soils that have distinct physical properties. These different soils interact with climate to produce characteristic growing environments for plants. Soils formed by the weathering of sandstone or granite tend to be coarse and sandy or gravelly, with relatively low water-holding capacity, while those weathered from limestones, shales, and basalts tend to be finer-textured, to contain more clay, and to have a relatively high water-holding capacity. In the mountains, loamy soils relatively high in clay definitely support more tree growth than sandy or gravelly soils. The explanation is that, once the root zone is filled with water, the rest of the water can drain out the bottom of the soil profile to join with subterranean groundwater. In a loamy soil, the root zone will contain more water when it is filled, making more water available to the trees that grow there, while in a sandy or gravelly soil, more water will be lost out the bottom. This is also true in the temperate environment of eastern North America, where sandy soils are observed to be more "droughty" than loamy soils.

Interestingly, however, in desert environments, the opposite phenomenon can be observed. Sandy soils definitely support more plant growth in deserts than loam or clay soils. The reason for this is that in dry environments, it rarely if ever rains enough to fill the soil profile to the point where water drains below the root zone. A much more important source of water loss from soil in deserts is through evaporation from the surface. And it turns out that sandy or gravelly soils lose much less water through surface evaporation than loam or clay soils, because once a coarse soil dries out on the surface, the below-ground reservoir of water is protected from further evaporation by the layer of dry surface material. In effect, the layer of dry sand acts as a mulch. Sandy soils have little ability to wick water to the surface, whereas clay soils continually wick water from below to replace that lost by evaporation, so that clay soils dry out evenly, and water stored deep below the surface is quickly lost. This means that more subsurface water will be available to plants over a longer period in sandy soil, a difference that is of considerable importance to desert plants.

Soil interacts with climate and topography in another essential way, and this involves feedback from the resulting plant community. These interactions influence fertility and organic matter (humus) content and indirectly affect the complex of microorganisms that occur in the soil and within the root zones of plants. It is easy to observe that dry desert environments support plant communities made up of mostly small plants at wide spacing, while wetter mountain environments support larger plants, such as trees, that grow in close proximity, often with a dense understory, as well. These contrasting scenarios represent differences in productivity, which in turn usually result in differences in the total mass of plant material per unit area that an environment can generate and sustain. Plant productivity generally increases as water availability increases and decreases at cooler temperatures.

The most productive plant communities in our region are the streamside and wetland communities at lower elevations, where warm temperatures and a longer growing season are combined with high water availability. Traditional home landscapes are usually managed to maintain high productivity, much like a wetland, and that is why they require a constant investment of resources such as water and fertilizer. This is in contrast to most native plant communities, which are made up of plants that are generally very frugal in their use of resources. For native plants, artificially rich topsoil and perpetual watering are definitely not necessary.

There is a straightforward relationship between plant productivity and the fertility and organic matter content of the soil, with more productive streamside, wetland, and mountain plant communities having a soil organic matter content substantially higher than that found in the soils of desert communities. But organic matter is a double-edged sword. For plants adapted to rich soils, life is good in such a soil. But disease organisms lurk in organic matter. Many desert plants, never having encountered such conditions, generally have little resistance to these diseases. Rich soil does not enhance their performance, and it may even threaten their lives.

This observation points to a more general paradigm: plants that come from different environments have different needs. Each combination of climate, topography, and soil creates a particular environment or habitat where a particular suite of native plants can be found growing together as a community. By matching plants to habitats that meet their requirements, and by grouping plants with similar requirements together, you can create designed landscapes made up of groups of plants that will thrive in each other's company.

With the help of this book, you can learn to recognize native plant communities without any special training in botany or ecology. You can learn to look at plant communities with the eyes of someone who wants to know what it is about being in a certain kind of place, surrounded by a certain community of plants, that feels so welcoming. If you have any doubt that the plant community itself is a major element of the magic of a wild place you love, no matter how spectacular the scenery, try imagining that place without any plants.

Fortunately, it is much more realistic to re-create the ambience of such a place in a designed landscape by using elements on the scale of plants and rocks than it is to attempt to re-create the scenery itself. If you observe closely, you will see how the feeling created by a place is evoked on many scales. These range from the grand scale of cliffs and rivers, which is frankly beyond reach in a designed landscape, to the miniature scale of the arrangement of pebbles, fallen leaves, grass clumps, and sand in a canyon bottom. And there are many scales in between. One of the most powerful ways to distill the essence of a place is to choose your scale carefully. You can also echo patterns observed on a grand scale, re-creating them on the more achievable scale of a designed landscape.

In order to assess what it is you like about a particular plant community and how you could go about capturing its essential features in a designed landscape, you will need a basic understanding of native plant community structure. In this book, we divide intermountain plant communities into four principal groups, based on climate and, more specifically, on average annual precipitation.

*Desert plant communities* occur where the average yearly precipitation (including both rainfall and the water contained in snowfall) is roughly ten inches or less. *Semi-desert communities* are found where precipitation averages more than ten inches but less than fifteen inches a year. *Foothill communities* occur where average annual precipitation is between fifteen and twenty inches, and *mountain communities* occur where annual precipitation averages more than twenty inches.

In the semi-arid climate that prevails in our region, average precipitation is somewhat of an abstraction, because there is tremendous variation in total precipitation from year to year. Being optimists, people tend to interpret above-average years as "normal" and label below-average years as drought years. But in fact, in most of the Intermountain West, more than half the years are below average. This is because wet years tend to be very wet, and to pull the average upward.

The average precipitation boundaries described above should not be interpreted as absolute, because many other factors, particularly evaporation rates, act to either accentuate or minimize the effects of too much or too little rainfall. Similarly, the boundaries between different plant communities are usually not sharp. Different communities often interfinger, weave, or blend, especially in situations where complex topography or geology creates a range of different microhabitats within an area. Another variable to consider is the time of year the precipitation is received. In the northern part of our region, summer rainfall is not common, even in the mountains, and rarely occurs at all in the valleys. As you go further south, the influence of subtropical summer monsoons becomes more pronounced, so that in southeastern Utah, for example, up to half the annual precipitation falls in the summer, even in the valleys.

Our concept of mountain communities for this book does not include subalpine and alpine communities, which occupy cool-summer habitats that are very difficult to emulate in landscapes at lower elevations; cultivation of these plants is best left to true aficionados. Similarly, our concept of desert communities does not include the warm deserts typical of the Southwest, which enter our region marginally in southwestern Utah and southern Nevada. Warm desert communities include many plants that are not cold-hardy in most of the Intermountain West. The mild winters in warm deserts are quite different from the cold winters characteristic of desert areas in most of our region.

We also do not include any formal treatment of wetland and streamside plant communities, as these generally require more water than most of us want to use on a landscape, though there may be special circumstances, such as the presence of surface water on the property, that would warrant the use of these plants. There is also merit to the idea of creating small areas in a designed landscape that provide limited habitat for high-water-use plants, as jewel-like oases in a landscape otherwise dominated by plants of drier places. Desert springs achieve much of their considerable impact through this kind of contrast with the surrounding countryside.

## Desert Plant Communities

Desert plant communities (average annual precipitation of ten inches or less) are characterized most of all by their open space. Like Zen gardens, the overall absence of plants in the desert is as important to the aesthetic effect as the placement of the plants that are there. Desert plants tend to be diminutive in size. The community is usually dominated by one to three shrub species, intermixed with small-scale patches and sprinklings of perennial grasses, succulents, and wildflowers. This is especially true on wide valley floors with fine soils and little topographic relief, where a few superbly well-adapted species, like shadscale, tend to dominate. And where the desert soils contain an excess of soluble salts, even fewer species can survive. At the very bottom of the valley floor, where runoff water collects and evaporates after storms, leaving behind its load of salt, the soils are too salty to support plants at all. These expanses of bare soil are called dry lakes, or playas.

Because of their lack of water, many desert valleys support little in the way of human habitation, and this undoubtedly adds to the sense of open space experienced there. But this stereotypic image of the desert as a vast expanse of little gray shrubs fading into the distance is only one of its facets. Where rocks outcrop on the surface, or where washes dissect the plains, there is often a dramatic increase in plant diversity, particularly in terms of wildflowers and succulents. And the added vertical relief, though often only on a small scale, greatly increases the visual impact of the plants that are there.

Because the plants in a desert are generally widely spaced, the color and texture of the ground, as well as the rocks, have a major impact on the aesthetic effect of desert communities. Many people find the red-orange sandstone and pale coral sand of the Colorado Plateau desert country immediately appealing. This warm background color brings out the blues and greens in the pale foliage of desert plants like prince's plume and Utah penstemon. The elephant-hide gray limestones of the Great Basin are less immediately accessible, but these rough rocks make a dramatic backdrop for desert sage, dwarf yucca, and desert needlegrass. And there are few scenes more striking than a desert garden of green Mormon tea, Indian paintbrush, and Indian ricegrass against a backdrop of columnar black lava rock. Curiously, rocky areas in the desert may actually have more available water than surrounding flatlands, because the bare

Sand dune community.

Mixed desert shrubland community.

Badlands community.

Rock outcrop community.

rock acts as a collector. The rainwater drains deeply through crevices and is stored below ground. This feature can be used to advantage in a designed desert landscape.

Dune areas in deserts also support uniquely beautiful plant communities. Many of these plants are hard to grow in cultivation unless your soil is sandy or otherwise very well drained. In traditional horticulture, sandy soil is considered a liability, but for native landscaping, it greatly increases the palette of plants that can be used successfully. Plants like sand sagebrush, with its fine, sea-green foliage and elegant, drooping form, are ideally suited to this environment.

Lastly, we cannot neglect to mention the unusual and beautiful array of species found on badlands in the desert. Badlands are areas that support minimal plant densities because their soils, usually gypsum or heavy clay, can pose serious problems for baby plants. These soils are found in an amazing assortment of colors, from chalk white to gray, maroon, ochre, and even mineral green. The plants found there are equally noteworthy. Lacy buckwheatbrush, prince's plume, and silver buckwheat are examples of plants most at home in this habitat. Fortunately, they do not seem to need such an extreme environment in order to prosper.

## Semi-desert Plant Communities

Probably the most typical and widely distributed plant community in the intermountain region is the semi-desert sagebrush steppe community (average annual precipitation of ten to fifteen inches). Its dominant species, big sagebrush, is a protean creature that also dominates some floodplain, desert shrubland, and mountain meadow communities, but in parts of the region with a semi-desert climate, it forms an association with bluebunch wheatgrass that could be called the signature community of the Intermountain West. Some other species typical of this plant community are rubber rabbitbrush, basin wildrye, Utah sweetvetch, sulfurflower buckwheat, and Lewis flax, but in sagebrush steppe that has not been heavily grazed or otherwise disturbed, the diversity of perennial grasses and wildflowers can be truly breathtaking.

The visual impression of a big sagebrush community could be described as rumpled. The shrubs themselves have an irregular form, so that from a distance, the community looks somewhat like a silver-green chenille bedspread. The black, stringy-barked trunks of the shrubs peek through the rumpled gray-greenery in a most characteristic fashion, an unmistakable look that many of those raised in sagebrush country love unabashedly. On the other hand, sagebrush has its detractors, as it tends to increase in the wake of abusive grazing—some people even label it with the highly inappropriate designation of "weed." The idea of using it in designed landscapes would greatly surprise this latter group. But even its worst detractors would have to admit that the smell of sagebrush after rain is elixir, evocative of everything that is wild and beautiful in the West.

Because more rainfall generally means more resources, sagebrush steppe communities are more productive than desert shrubland communities, and this translates to higher plant density and less bare ground. The ratio of shrubs to understory species in this community is extremely variable and dynamic. In pre-settlement times, this shifting relationship was regulated by natural wildfire.

Sagebrush is killed outright by fire, whereas most of the perennial understory grasses and wildflowers are fire tolerant, sprouting back readily following a burn. The tendency of the sagebrush to increase gradually at the expense of the understory species was thus kept in check, but these fires were very infrequent, perhaps on the order of every fifty to a hundred years, and usually not very large. The result was a shifting mosaic of patches with varying sagebrush density and dominance, and with very high plant diversity on a landscape scale. In a designed landscape, it is probably best to start with relatively low sagebrush density, so that the shrubs provide an attractive structural backdrop for what is essentially a meadow of grasses and wildflowers. The composition of the meadow will shift with time, and the sagebrush will gradually increase, but it will be many decades before it reaches a level that strongly impacts the understory species.

In the northeastern section of our region, a plant community called palouse prairie used to cover vast areas in the semi-arid and lower foothill precipitation zones. Its species composition is much like that of sagebrush steppe, except without shrub dominance. Palouse prairie is found on deep, fertile loess soils, which are derived from glacial sediments. These are among the best agricultural soils on earth, so it is no surprise that palouse prairie is an endangered ecosystem, currently occupying less than one percent of its former range. Most of the palouse prairie country is now farmland.

Palouse prairie is a true grassland, like the prairies of the Great Plains, but differs in that its principal grass species are cool-season bunchgrasses, not warm-season grasses like big bluestem and Indian grass, which supported huge herds of bison in pre-settlement times. Because they did not evolve with large herding ungulates like bison, the grasses of the palouse prairie and sagebrush steppe are not especially grazing-tolerant. And because the palouse prairie environment has dry summers, most of its species are either spring or fall flowering. This is in contrast to the prairies of the Great Plains, which receive much of their moisture in summer and are home to a host of spectacular, summer-flowering species, many of which have entered the nursery trade.

Strictly speaking, the palouse prairie historically was restricted to the area in northeastern Washington and northwestern Idaho north of the Snake River, but ungrazed sagebrush steppe where sagebrush and other shrubs have been temporarily removed by wildfire or other disturbance bears a strong resemblance to true palouse prairie. For this reason, we often encounter this term in reference to areas outside the natural range of this community type. Palouse prairie probably represents the best natural model for a wildflower meadow community under the semi-arid, winter rainfall climatic conditions characteristic of most of the urban areas of the Intermountain West. But summers are cooler in the palouse prairie than they are in the sagebrush steppe areas to the south. This

Sagebrush steppe community.

Palouse prairie community.

Mixed semi-desert shrubland community.

factor needs to be taken into account when designing palouse prairie communities, which may need occasional supplemental water in hotter parts of the semi-desert zone.

Sagebrush steppe and especially palouse prairie communities reach their maximum expression on deep, fertile soils developed from wind- or water-deposited sediments. Where soils are shallow, or derived directly from underlying rocks, as in much of the Colorado Plateau, the semi-desert plant community is often made up of a diverse mixture of shrubs and small trees, with big sagebrush itself playing a relatively minor role. If you picture a typical landscape in Arches, Canyonlands, or Capitol Reef national parks, you will likely be picturing this kind of semi-desert plant community. Pinyon pine and Utah juniper, more characteristic of slightly wetter areas over most of their range, move down into the semi-arid zone in places where they can root deeply into rimrock sandstones and shales and take advantage of the stored water there. Statuesque tall shrubs and small trees like cliffrose, singleleaf ash, oakleaf sumac, Utah serviceberry, and New Mexico privet also occur in this zone, especially in the rocks or along dry washes where they can capture runoff water after storms. Typical understory species may include many of the same small shrubs found in desert communities, such as lacy buckwheatbrush, sand sagebrush, and green Mormon tea. Handsome shrubs like Apache plume, Fremont barberry, and littleleaf mountain mahogany are common, along with succulents like dwarf and datil yuccas. Perennial grasses and wildflowers are also abundant and sometimes very colorful in this community, though they are usually widely spaced.

Although these semi-desert communities also receive ten to fifteen inches of precipitation annually and are potentially as productive as sagebrush steppe and palouse prairie communities, low soil fertility frequently limits productivity, and plant cover is often not much higher than the plant cover found in deserts, although individual shrubs and trees are much larger. These soils tend to be infertile because they are formed directly from the single kind of rock that lies beneath them. A single type of rock will almost always generate a soil that has some kind of nutrient deficiency, whereas wind- and water-deposited soils are made up of mixtures of rocks that are more likely to contain a full complement of nutrients. As in desert communities, the color and texture of the rocks and the ground are major components of the aesthetic effect for semi-desert communities growing on soils developed directly from underlying rocks.

## Foothill Plant Communities

In the foothill communities of the intermountain region, the climate (fifteen to twenty inches of average annual precipitation) is wet enough to support trees, though many of these little trees would likely be called by some derogatory name like "scrub" in comparison with the trees we associate with the forests of wetter climes. For example, Gambel oak, a community dominant in the foothill zone and one of the few native oaks of the region, is often called "scrub oak," and it is true that even at its most magnificent, it might only reach forty feet in height. Gambel oak or other small trees, along with their associated understory shrubs, grasses, and wildflowers, form communities called foothill woodland, mountain brush, and mountain mahogany woodland. Each is characteristic of a particular set of conditions within the context of the foothill precipitation zone.

The foothill woodland community is usually found at the drier end of the foothill zone, often on shallow soils derived directly from underlying rock. It is best developed in the southern half of the region. This community is usually the least productive of the foothill communities because of lower precipitation, low fertility, and thin soils. Considerable bare ground is usually in evidence. The dominant tree species are pinyon pine and various species of juniper, but foothill woodland also includes a wealth of other shrubs and small trees, which may sometimes be very abundant. On the Colorado Plateau, foothill woodland includes many of the same species as semi-desert shrubland, including singleleaf ash, green Mormon tea, Fremont barberry and dwarf yucca. This is the community where cliffrose, littleleaf mockorange, squawapple, and other flowering shrubs put on prominent displays, each in its season. It also supports an abundance of showy perennials, including Palmer, firecracker, Utah, dusty, and Bridges penstemons, sundancer daisy, Hopi blanketflower, and desert four o'clock. As in the other plant communities where productivity is relatively low and vegetation structure is open, the aesthetic of foothill woodland communities

Foothill woodland community.

Mountain brush community.

Mountain mahogany woodland community.

is often defined as much by the color and texture of the ground and rocks as by the plants themselves.

The mountain brush community is usually found on somewhat higher-elevation sites or on northern exposures within the foothill zone, where temperatures are cooler. Mountain brush is the most productive of the foothill communities, often with dense plant cover, especially on more favorable exposures. It usually occurs on relatively deep, fertile soils. It varies considerably in tree composition from south to north in the region. In the south, Gambel oak and bigtooth maple are usually the dominant overstory species, while further north, you are likely to find chokecherry, serviceberry, mountain ash, netleaf hackberry, and other small trees. Most of these trees have leaves that change color in autumn, and spectacular fall color is one of the hallmarks of this community. But evergreen trees like Rocky Mountain juniper may also be abundant. Many mountain brush dominants, including Gambel oak and bigtooth maple, are clump-forming species capable of sprouting back after burning, so that the size of the trees is often a reflection of how long it has been since the last wildfire. But these trees do grow larger in more favorable canyon environments than they do on open slopes, indicating a positive response to a little extra water.

The mountain brush community often interfingers with the sagebrush steppe community at its lower edge, and these two communities share many understory species, including oakleaf sumac, shining muttongrass, basin wildrye, bluebunch wheatgrass, Utah sweetvetch, and Wasatch penstemon. Mountain brush also shares understory species, such as creeping Oregon grape, mallowleaf ninebark, and mountain snowberry, with adjacent mountain communities.

Mountain mahogany woodland occupies a special niche at the upper end of the foothill precipitation zone. It usually occurs on high, windy ridges, on very shallow soils over bedrock. It is dominated by a single species, curlleaf mountain mahogany, a small but beautiful broadleaf evergreen tree, and usually has a very sparse understory because of the rocky soil. The community also has low productivity because of this nonexistent soil, and the tree uses its evergreen leaves to hang on for as long as possible to what nutrients it has garnered. Curlleaf mahogany is not fire tolerant, but because of the open nature of this community, stands rarely burn, and the slow-growing trees become more twisted and full of character as they age.

## Mountain Plant Communities

In the Intermountain West, true forests occur where total annual precipitation exceeds twenty inches. At the higher elevations where these forest communities occur, summers are substantially cooler than in the valleys below, and a significant fraction of the precipitation comes in the summer as rain. In a few favored locations, such as certain ski resorts east of Salt Lake City, annual precipitation may be as high as fifty inches. But the plant communities in these favored locations are no more water loving than those that occur where the precipitation is only thirty inches. The reason for this is that essentially all the extra precipitation comes in the form of winter snowfall. Once the soil root zone is full, as mentioned before, the rest of the snowmelt either runs off the surface or sinks below the level of the plant roots and becomes part of the groundwater. This means that the reservoir of snowmelt water available to plants cannot increase beyond that which the soil can hold. This is the soil water reservoir that is present at the beginning of the growing season, and it is often more a function of soil type and slope than of total snowfall, once snowfall has been sufficient to fill the root zone.

At least as important as this snowmelt reservoir, especially for shallow-rooted understory plants, is rainfall that arrives during the growing season. Throughout much of the southern part of our region, this totals approximately six to eight inches, which is usually enough to support a lush understory. Summer rainfall is lower in the northern part of the region, farther from the effects of summer monsoonal storm systems, and understory productivity tends to be correspondingly reduced. These two sources of water need to be taken into account when designing landscapes based on forest communities. Even if winter precipitation is sufficient to provide the total annual water requirement for trees, it will still be necessary to water the trees as well as the understory if the summer is dry.

The species composition of mountain communities is highly variable, and the different communities commonly intergrade and blend. The mountain zone, as we define it—that is, excluding subalpine and alpine areas—includes four characteristic plant communities: ponderosa pine parkland, aspen forest, mountain evergreen forest, and mountain meadow.

Ponderosa pine parkland community.

Aspen forest community.

Ponderosa pine parkland usually occurs on soils derived from granite or sandstone, which tend to be relatively shallow, coarse, somewhat acidic, and nutrient poor. Because of these poor soils, ponderosa pine parkland is probably the least productive of the mountain communities. It usually consists of an open, parklike stand of ponderosa pine with a relatively sparse understory dominated by bunchgrasses such as bluebunch wheatgrass, Indian ricegrass, shining muttongrass, and occasionally blue grama or little bluestem in the south. Some bare ground is usually visible, though a mulch of pine straw tends to cover the spaces between the plants. Other understory species characteristic of this community include creeping Oregon grape, mountain lilac, alderleaf mountain mahogany, rosy pussytoes, butterfly milkweed, and sundrops. Ponderosa pine parkland is a fire-resilient community in its pristine state. Frequent low-intensity ground fires prevent the establishment of most tree seedlings but do not damage larger trees, and most understory species resprout readily. This is how fire maintains the park-like structure.

Quaking aspen forest may be the most well-recognized and beloved intermountain plant community. Aspen, the dominant overstory species, is an elegant little tree reminiscent of eastern paper birch, and its dappled shade supports a rich array of understory species. Aspen forest is usually found on more fertile, slightly alkaline soils weathered from basalt or limestone, and it is generally a much more productive community than ponderosa pine parkland, though this varies with soil depth, elevation, and exposure. Common shrub associates include mountain snowberry, mountain lover, common juniper, and creeping Oregon grape. Aspen forest is home to a host of attractive wildflowers, including Rocky Mountain and western columbines, blooming sally, sticky geranium, tall larkspur, and Wasatch penstemon. The understory of an aspen forest usually covers the ground completely.

Aspen is a clump-forming species that sprouts back after fire. It is not especially shade tolerant, and it tends to attain dominance only in the wake of forest fires that temporarily remove white fir and other conifers in the mountain evergreen forest. Eventually the evergreen trees recolonize the aspen forest, and it is common to see aspen communities filling up with adolescent conifers. Without fire, it is likely that most of the aspen forest would eventually disappear, swallowed up in the shade of these adolescent conifers grown large.

Evergreen forest community.

Mountain evergreen forest occupies essentially the same kinds of habitats as aspen forest, and, as explained above, the length of time since the last fire determines which community will be present. White fir is usually the dominant tree species in the southern half of our region, along with Douglas fir and sometimes blue spruce. Farther north, lodgepole pine and western white pine can play major roles. The understory mostly includes a more shade-tolerant subset of the species found in aspen forest, including mountain lover and common juniper, as well as columbines. But in general the understory is much less diverse, and as the trees age and the canopy closes, the ground beneath can become almost barren of plants.

Mountain meadow community.

In contrast to the mountain forest plant communities, where the trees create varying degrees of shade for the plants that grow beneath their canopies, the mountain meadow community usually has no trees, and this brightly lit environment creates habitat for many sun-loving species. Big sagebrush reappears as a dominant in this environment, this time in its mountain big sagebrush form. Mountain snowberry and many other shrubs that grow beneath the aspens are also found in the full-sun meadow habitat, as are many of the grasses and wildflowers. And there are many additional species that usually shun the shade of the aspens but thrive in the open meadows. These include prairie smoke, showy daisy, and meadow fire. Because of their high productivity, there is usually little bare ground to be seen in mountain meadow communities.

*Silver buckwheat*

# Chapter Two

# How to Design Native Landscapes

Now that you have completed your tour of native landscapes, imagine some possibilities for taking a new approach to designing landscapes in the Intermountain West. What might be some characteristics of these landscapes? First, they would be experientially rich. Their spatial character would incorporate a sense of mystery or intrigue, making us want to explore them further. The forms, colors and textures of the plants would be harmonious, just as the forms, colors, and textures of plants growing together in the natural landscapes of the Intermountain West exhibit harmony.

Second, these designed landscapes would be ecologically sound. Plants would be matched both to the regional environment and to the microhabitats in which they are placed. Because of this matching, the need for supplemental water would be reduced. And because there would be few, if any, areas of mowed turf, there would be little need to use fossil fuels to mow them. These landscapes would not include invasive introduced species that have the capacity to escape into the native landscape and reduce the natural diversity there.

Third, these landscapes would be "of the place". The character of the designed landscapes would draw on the rich menu of regionally distinct landscapes. These designed landscapes may or may not be naturalistic in form. They may very well be artistic distillations of native plant communities of the Intermountain West. But because they draw on the plant species and patterns of the region, they would speak unmistakably of the place.

Finally, these landscapes would be dynamic, changing over time. They would exhibit the rich change in color through the seasons that we see in the natural landscapes of the Intermountain West. And there would be other, longer-range changes as well, such as those resulting from the reproduction and spread of some species, or the phasing out of some species and the phasing in of others as the amount of shade increases under expanding tree canopies. Hence, there would be new things to discover in the landscape over the years. A whole new concept of landscape management would apply: one that is not oriented toward "freezing" the landscape in time, but instead guides the direction and rate of change.

*Mystery*: What's around the bend?

## Getting Started on Design

How do you go about designing landscapes that would have these characteristics? What do you need to know? What are the logical steps you might go through? One starting point, certainly, would be to familiarize yourself with the distinctive aesthetic characteristics of the natural landscapes in a variety of habitats in the Intermountain West. In the end, you will not necessarily be copying them in design, but trying to capture their essence. Two environmental psychologists in the School of Natural Resources at the University of Michigan, Drs. Steve and Rachel Kaplan, have identified four characteristics that occur frequently in natural landscapes, ones to which people seem to have a positive response. These are: *mystery, complexity, coherence,* and *legibility*. We will discuss how these features are exhibited in western landscapes, and we will also characterize the prevalent lines, forms, colors, textures, and patterns in representative intermountain habitats. These could provide a vocabulary for designing landscapes on sites with particular sets of environmental characteristics.

*Complexity*: The varied shapes and textures of the plants in this desert sand garden are set off by the bare rock around them.

*Mystery* is the quality of having part of the landscape concealed, thus leaving more to be discovered. A meandering stream typically provides the element of mystery as it bends out of sight beyond peninsulas of rocks or vegetation. Similarly, open, riverlike spaces—some broad and some narrow—can occur on dry land, where clumps of trees or shrubs partially block the view. Mystery is also

*Coherence*: In nature, individuals of each plant species cluster in the areas best suited for them.

*Legibility*: You can see how to get into the meadow by going through the "doorway" between the trees.

found in layers of one mountain range behind another, then behind another, luring your eyes forward until they fade into the distance.

*Complexity* implies the qualities of diversity and intricacy in line, form, color and texture. In the natural landscape, this property is manifested by a diversity of plant species. It is not uncommon in nature to find at least six or eight plant species growing together within a square yard, and in the mountain meadows of high elevations, this number may be doubled or tripled. This leads to the possibility of always having more to discover as well, and a diversity of flowering at different times. Contrast this with the monocultures we have come to expect in designed-and-managed landscapes, as in lawns and single-species ground cover plantings.

*Coherence* is reflected in the fact that the complexity in natural landscapes is balanced by a logical order in the distribution of species diversity. There is patterning, with individuals of a particular species showing various degrees of aggregation, or gathering together, based both on the way plants reproduce and interact and on minor differences in soils, moisture, or light. Pattern is perhaps most obvious in open landscapes such as marshes and salt flats, where vegetation is clearly grouped according to minor differences in water depth or salinity.

*Legibility* refers to our being able to read how we can move through the landscape, and thus is related to its spatial configuration. Again, rivers or river-like spaces, which are visual pathways through the landscape, provide legibility. The most basic form of legibility is space, and in the Intermountain West, there is space in abundance, both on a landscape scale and on the scale of individual plants.

As mentioned earlier, a central source of inspiration for designing landscapes that have a distinctive Intermountain West aesthetic is the natural landscape of the region, and the plant communities that occupy different portions of the landscape. These may be sharply separated or, more often, grade or feather gradually from one to another. In observing such communities, we will be looking for the ecological and aesthetic characteristics that can be translated into designs for landscapes that capture their essential qualities.

# Design Features of Native Landscapes

## Species Composition

A primary feature of a natural plant community is its species composition. Rarely would we have the opportunity to replicate a native community in all its species richness, but we should be especially aware of the following categories in the plant community that is our source of inspiration:

*Dominant species*: These are typically the biggest plants in a particular community, namely, the canopy trees in a forest community, the shrubs in a shrub steppe, or the tallest grasses in a grassland community. The dominant species constitute much of the visible mass in the mass/space composition of the community. They are also influential from an ecological perspective, in that they frequently shade or otherwise compete with smaller plants, and often drop leaves which contribute to the organic matter and water-retaining capacity of the soil.

*Abundant species*: The plants that may be small but that occur in the largest numbers within a community are important as well, both ecologically and aesthetically. They may be trees, shrubs, grasses or perennials. Because of their abundance in a natural community, they logically have a place in a designed landscape that is inspired by that community.

*Visual essence species*: These species are important to the landscape, not because they are necessarily dominant or abundant, but because they are uniquely representative and almost symbolic of a particular environment. They may be indicator plants that only occur within a particular type of community. They could be dominant species that are visually important, such as the spire-like conifers of mountain forest communities. Or they may include abundant species such as quaking aspens, which are visual essence species because of their shimmering leaves, especially in the fall when they turn yellow to gold in contrast with the dark conifers. But the visual essence species of this community can include diminutive plants, as well; the delicate Rocky Mountain columbine, for example, adds a distinct flavor to this landscape.

## Spatial Configuration

Just as it is important to observe species composition, it is also useful to note how the plants are spatially distributed in a wild plant community. Different plant species exhibit different densities and patterns of distribution. One possible arrangement is truly random, where there is no perceptible pattern, with some individuals of the species more widely spaced and others closer together. A second possibility is for a species to be regularly distributed. We may see this phenomenon in very arid situations, where individuals of a species are spaced at

| | Dominant | Abundant | Visual Essence |
|---|---|---|---|
| **Mountain** Aspen Forest Community |  *Quaking Aspen* | 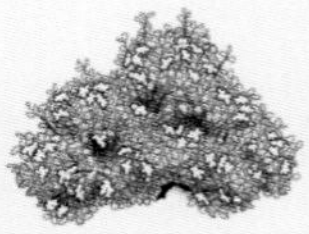 *Mountain Snowberry* |  *Rocky Mt. Columbine* |
| **Foothill** Mountain Brush Community |  *Gambel Oak* |  *Creeping Oregon Grape* |  *Maple Mallow* |
| **Semi-desert** Sagebrush Steppe Community |  *Big Sagebrush* |  *Bluebunch Wheatgrass* |  *Sulfurflower Buckwheat* |
| **Desert** Sand Dune Community |  *Sand Sagebrush* |  *Indian Ricegrass* |  *Dwarf Yucca* |

Examples of dominant, abundant, and visual essence species for four representative Intermountain West plant communities. The dominant species are generally the largest and most conspicuous, while the abundant species are comprised of numerous individuals and may cover large areas. The visual essence species are highly representative of a community type, and usually have notable features that make them stand out.

nearly equal intervals, probably because each individual preempts the available water within a certain radius.

The majority of plant species are not distributed in either a truly random or a totally regular pattern, but occur with some level of aggregation, or grouping. This is very obvious with species like quaking aspen, which tend to spread clonally, with young shoots occurring at the edges of ever-larger patches, until they reach a barrier or a limitation of resources. Other plants may be aggregated not because of their reproductive strategy, but because they have specific microhabitat needs, and hence are gathered together where those needs are best met.

A particular type of aggregation pattern that seems to occur frequently in natural plant communities is called a drift. This occurs where there is a concentration of a particular species which then thins out at the edges, with smaller, often younger individuals trailing off the central aggregation. This pattern can be readily adapted in a planned design, with drifts of different species interacting in a directional flow across the landscape.

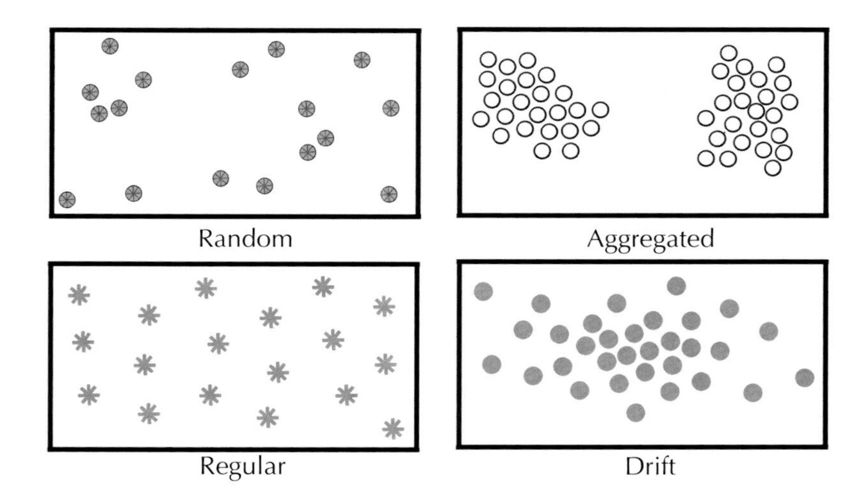


Individuals of the same species within a plant community may be randomly or nearly regularly spaced, or they may show different degrees of aggregation. A common grouping pattern, termed a *drift*, has individuals more closely spaced at the center of the group and trailing away at the edges.

## Structure

Another important characteristic of natural plant communities is structure, namely the proportion of the space that is occupied by plants, and by plants of different heights and growth forms. Structure is related to productivity, as we discussed earlier. Intermountain West landscapes range from riparian gallery forest and mountain forest communities with high tree canopy cover, to the mountain brush community that typically has moderate tree canopy cover, to foothill woodland communities that usually have 50 percent tree canopy cover or less, to the desert landscape with no tree canopy cover at all. Shrubs also contribute to the structure of a plant community, creating mass at eye level or below that contrasts with the more open space formed by low-growing herbaceous vegetation.

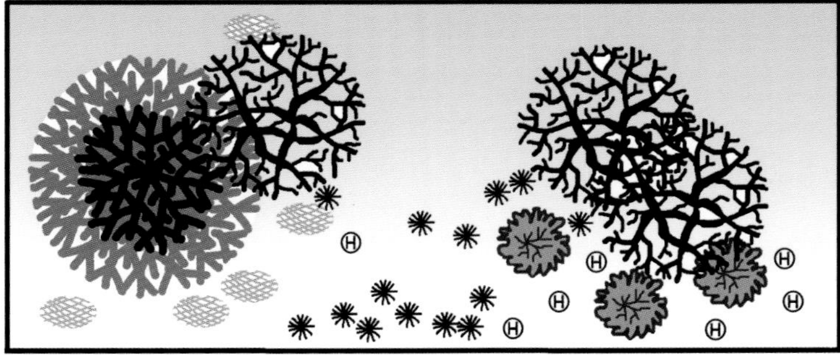

As an example of the way structure influences aesthetic character, consider the mountain brush community, common in the foothills where much of the urban development is occurring. The spatial structure of this community characteristically possesses a pleasing amount of both mystery and legibility, with grassy spaces flowing between islands and peninsulas of small trees and shrubs.

## Aesthetic Elements

In any natural landscape, certain aesthetic elements are present: lines, forms, colors, and textures. In our study of natural plant communities as models for designed landscapes, it can be useful to identify the elements that make them distinctive. Then, in designing landscapes in environments that match those of the natural communities, we are better prepared to capture their aesthetic essence.

Lines occur in many different aspects of the landscape. There are the lines where land meets water, such as along the edge of a river or lake, and those where land meets sky. In the Intermountain region, these lines range from the broadly horizontal sweep of a desert or a salt flat to the gently to steeply ascending lines defining mountain slopes. Lines occur in the vegetation of plant communities, as well. The trunks and branches of trees, especially those of deciduous trees in winter, are lines against the sky. They may be arching lines, ascending vertical lines, or irregular, angular lines. Lines occur in the narrow leaves of grasses and grass-like plants, often arching or in fountain-like clusters, and in the foliage of desert plants such as yuccas and green Mormon tea.

Forms include the three-dimensional shapes of the plants: the narrowly conical shapes of conifers in high-elevation landscapes; the irregularly rounded contours of pinyon pines and junipers set against fountain-like clumps of grasses and the low, mounding forms of sagebrush and other shrubs; or the consistently low mounds or billows of shrubs that blanket the shrub steppe community. Form, of course, also includes landforms and the shapes of rocks, both critical to the unique character of the Intermountain West.

Color is one of the most important aesthetic elements, clearly setting Intermountain West landscapes apart from those of the rest of the country. These include the colors of rock, which vary within the region, of exposed soil, and of vegetation. Eastern landscapes (as well as intensively irrigated western landscapes) tend to be overwhelmingly green. The native landscape of the Intermountain West, on the other hand, has a distinctively different color tone. Particularly characteristic is the prevalence of silvery green tones at the shrub level in many native regional landscapes. This is attributable to the abundance of sagebrush, as well as rabbitbrush, shadscale, and other drought-hardy shrubs. Tan to copper-colored grasses contrast with the silvery shrubs. Especially in the foothill communities, the warm, light tones of grasses contrast with the woody vegetation. In the foothill woodland association, they stand out against the dark muted greens of pinyon pine and Utah juniper. In the mountain brush community, they contrast with the colonies of bright green Gambel oaks and bigtooth maples. That community has the added attribute of having patches of fiery red in October, when bigtooth maples and sumacs take on their fall colors.

At higher elevations, a key color characteristic is the stark contrast between the blackish-green conifers and the light-colored aspen leaves, which are warm green in summer and yellow to gold with tinges of orange in the fall. Also distinctive in meadow openings at high elevations are the jewel-like colors of such

### Exercise Your Vision

Look at this scene the way an artist might.

*Lines:* The tree trunks form vertical lines; the distant ground forms horizontals. What other lines can you find?

*Forms:* Notice how the rounded shapes of the shrubs are repeated. How are the tree shapes different from the shrub shapes?

*Colors:* The warm, sunny colors of the vegetation contrast with the cool shadows. If you were taking color notes for this scene, what would you name the various colors? How would you distinguish the green of the shrub in the lower left from the green of the yucca behind it?

*Textures:* See how the rumpled texture of the foreground shrubs contrasts with the fine, twiggy textures behind them. Even the snow on the ground has a texture. Compare the texture of the snowy foreground to the distant rocks.

Finally, ask yourself how you feel about this place. Is it a place you would like to be? What do you like or dislike about it? If you were painting a picture of it, would you change anything?

flowers as Rocky Mountain columbine, sticky geranium, Wasatch penstemon, showy daisy, and meadow fire, set amidst the grasses.

Texture is also of key importance in the aesthetic of native communities. In arid climates, leaves are often diminutive in size, as an adaptation to avoid excessive water loss. As a result, there is an abundance of fine textures in the landscapes of the Intermountain West, for example, the above-mentioned grasses and silver-gray shrubs that form a background matrix of fine lines. Conifers, whether pinyon pines and junipers in lower elevations or spruces and firs at high elevations, with their narrow needle-like leaves, are also fine-textured. Quaking aspens, with their small fluttering leaves, have a fine texture, and even Gambel oak and bigtooth maple leaves appear relatively fine from a distance, due to the "teeth", or indentations, in the leaves. This abundance of fine-textured foliage is one reason why adding rocks, which are coarse textured and bold in their aesthetic effect, creates such a powerful contrast.

The aesthetic essence of a plant community is not easily quantifiable, and may seem elusive. Rather than trying to objectively calculate the aesthetic component in the field, one approach is to sketch or do watercolor interpretations of the landscape. Pencil or pen-and-ink sketching is particularly useful for grasping the essential lines and forms in the landscape. Watercolor can be very helpful in observing and capturing the subtleties of landscape color. Even if you are not an accomplished artist, the very process of observing a landscape closely enough to render its essential lines, forms, colors, and textures on paper can provide lasting impressions which will be useful as you proceed to design a regionally appropriate landscape. Photography can also be helpful as a way to record the aesthetic characteristics of natural landscapes, to use as a reference during the design phase.

Another useful exercise is to write down your observations and feelings about a particular native landscape, which will help you connect to the essence of the landscape. Ask yourself the following kinds of questions when you are experiencing a native landscape that you find appealing: What is it about this native landscape that I find attractive? What is the most noticeable or striking thing about the landscape? What are the predominant colors and forms? How are the spatial and structural themes repeated at various scales? Are there important elements that are not so much visual as kinesthetic, or perceived by other senses, such as the rough feel of netleaf hackberry bark or the smell of cliffrose foliage in the afternoon sun? By making a written record of the answers to these sorts of questions, and thinking about how you could create a landscape that evokes similar pleasant feelings, you have accomplished a lot of the preliminary work needed to create a native landscape design.

## Step-by-Step Native Landscape Design

Now that you have some ideas for the aesthetic effect you would like to achieve in your native plant landscape, it is time to settle down to the actual work of creating the design. This process has several steps: characterizing your site, considering how you will use your landscape, planning the landscape watering zones, creating the mass/space diagram, laying out the landscape plan, selecting the plants, and preparing the planting plan. Each of these steps involves feedback from the other steps, so it is a good idea to read through the description of the entire process before trying to apply what you have learned.

**Inspiration from Natural Landscapes**

Color studies by four different individuals based on an outing to the Wellsville Mountains near Logan, Utah, in autumn. A photo of the scene is shown for comparison. Color studies like these can form the basis for the beginnings of design work using natural landscapes and native plants communities as inspiration.

## Characterizing Your Site

The first step in the design process is to spend some serious time investigating your site. Perhaps the best place to start is to consider your site in the context of the place where it is located. This includes its geographic location, elevation, and macroclimate—in other words, the habitat it will provide for the native plants you plan to include in your landscape.

A critical next step in the design process is to take an inventory of existing conditions on and around the site itself, starting with the preparation of a base map. This base map is the framework upon which the remaining design steps are developed. The base map should illustrate the major existing features while remaining uncomplicated and basic. It should be drawn to scale, with the scale and cardinal directions indicated. It will be useful to make several photocopies of the base map. Use an oversized piece of paper if you can—there will be a lot to record. Once the base map is obtained or prepared and copied, use a copy to record the information you collect during the inventory you make on your next hike around your site.

*Property Lines*: This information may be contained on the house plan or deed statement, or it may be obtained from the county assessor.

*Existing Building Outlines*: Include notes on the heights of buildings, because height will affect the extent of cast shadows. Window and door locations should also be identified on the building outlines, because they will affect foot traffic and viewing patterns.

*Existing Hardscape Features*: Hardscape includes driveways, walkways, and any other paved or finished surfaces or features, such as walls, fences, decks, and patios. You may also include features such as outside lighting, air conditioning units, and gutter downspouts.

*Utility Lines and Easements*: Indicate the location of overhead and underground utility lines, including cables, along with information on easements and setbacks, as well as the location of water and electric meters. The location of underground lines can be obtained from Blue Stakes or similar community services that provide utility line marking. Easements and setbacks can usually be found on a house plan or plat map.

*Site Position*: The site position relative to the surrounding area may suggest possible opportunities and constraints for the design. Adjacent features could be things that you want to emphasize in the design (such as a good view) or deemphasize (such as the side of your neighbor's house). Neighboring land uses may create microclimate effects, such as shading. Paved roads may create hot conditions and increased salt in the adjacent soil. Views, good and bad, both within the site and off-site, should be noted on the base map.

## How to Get the Base Map

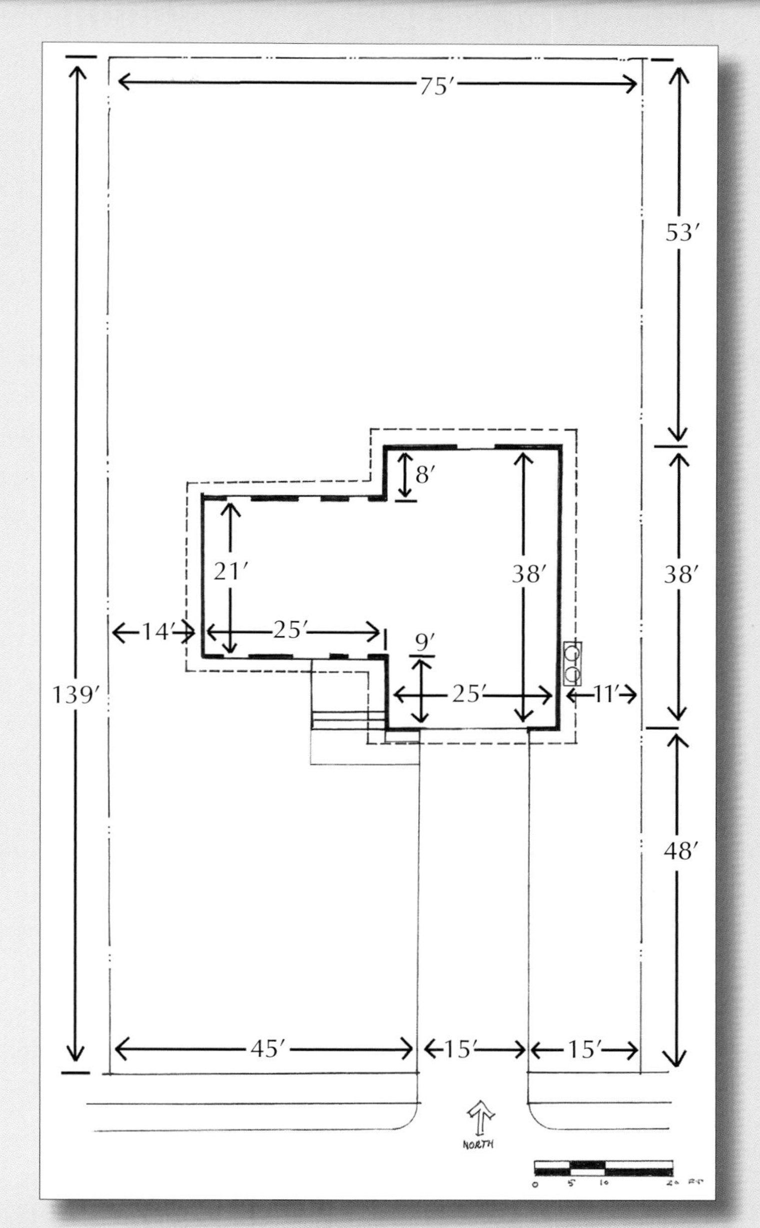


The base map is a scale diagram of your property and the outside dimensions of your house, with the scale and cardinal directions marked. There are four ways to obtain a base map of your property:

- If you have the paperwork from the purchase of your house, you should have a scaled base map tucked away with your deed.
- You can go to the county assessor's office, which has legal descriptions of all properties in your county. Someone there can help you obtain a copy of your plat map or house plan for a small fee.
- You can create your own base map by doing the measurements yourself. You will need a long measuring tape or measuring wheel and at least one assistant to record the measurements. First measure the boundaries, then the distance from the house to one boundary, then the dimensions of the house. The measurements can be written on a rough map, and the map can be redrawn to scale later. You will also need a compass to determine which direction your house faces on the property.
- You can have your property professionally surveyed. This might be worth it if the property is large, if the house has a complex shape, or if there is a lot of topographic relief.

*Topography, Slope, Aspect, and Drainage*: If your site has areas that are sloping, make note of the steepness and orientation (aspect) of the slopes. If your site has a lot of topographic relief, a good approach is to sketch contour lines onto the base map. Contour lines connect points that have roughly the same elevation, so that hills and hollows are apparent as roughly concentric sets of closed lines. Be sure to note whether the spot inside the innermost closed line is a hill or a hollow. These relationships are important, because south- and west-facing slopes tend to be hotter and drier than north- and east-facing slopes. North slopes will hold snow the longest. Similarly, the north side of a building will be cooler and shadier than the south side. The lay of the land is also important in determining drainage patterns on the site, which can be an important element when designing with native plants. By recognizing where the water discharges from the roof and where it drains onsite, you can utilize these areas for higher-water-use plants. This makes better use of the water than installing structures to drain it away offsite.

*Soils*: Understanding the characteristics of the soil on your site is important to the success of a native plant landscape. The best way to begin is to get down and dirty, dig some holes, and carry out a few Johnny-on-the-spot soil tests. Soon you will know whether your soil is sandy or heavy, whether it contains gravel or cobbles in the subsoil, whether it has an organic surface layer and, if so, how thick, and whether it has a hardpan (relatively impermeable layer) near the surface. You will also find out if you have compaction problems as a result of construction equipment or other heavy traffic in the past. It is important to check your soil in more than one spot. Often topsoil is applied after construction, to different depths depending on the distance to the house, so one part of the property may have a relatively deep organic soil while another part might have a very thin soil over a gravelly or cobbly subsoil. This soil assessment will give you an idea of which parts of your site are suited to different kinds of plants. If your soil is deep and highly organic, you will do best with plants from the foothill and mountain zones, where such soils occur naturally. Conversely, if your soil is rocky and "poor," that is, low in fertility, it may be admirably suited to plants from the desert and semi-desert zones.

One important soil property that is not easily observed on the spot is salinity, though a white, powdery deposit on the surface of drying soil is likely to indicate this problem. High salinity is generally confined to heavy bottomland soils in semi-desert or desert environments. If this fits your site description, you should have your soil tested to determine whether it is saline. High salinity can also be a problem locally, where de-icing salts are used in excess.

*Existing Landscaping*: Evaluating the existing vegetation on the site is important for several reasons. This process will be very different depending on the age

## Characterizing Your Site

Use this checklist to make sure you record all the information you need about your site:

- Location (latitude, longitude) and elevation—quick ways to obtain this are a GPS unit or a map website (see resources)
- Climate, including mean annual precipitation—use online climate websites (see resources)
- Plant cold-hardiness zone—call your state's extension service or use online resources
- Natural plant communities that occur in the area
- Existing hardscape features
- Utility lines and easements—call Blue Stakes or another utility marking service for the locations of underground lines
- Site position and adjacent features
- Views—good and bad
- Slope and aspect
- Topography and drainage
- Soil features—texture, drainage, depth, organic matter
- Existing landscaping—evaluate whether to keep or remove it
- Current watering system—location of hose bibbs/sprinkler system
- Existing weed problems—weed identification and seriousness
- Microhabitats

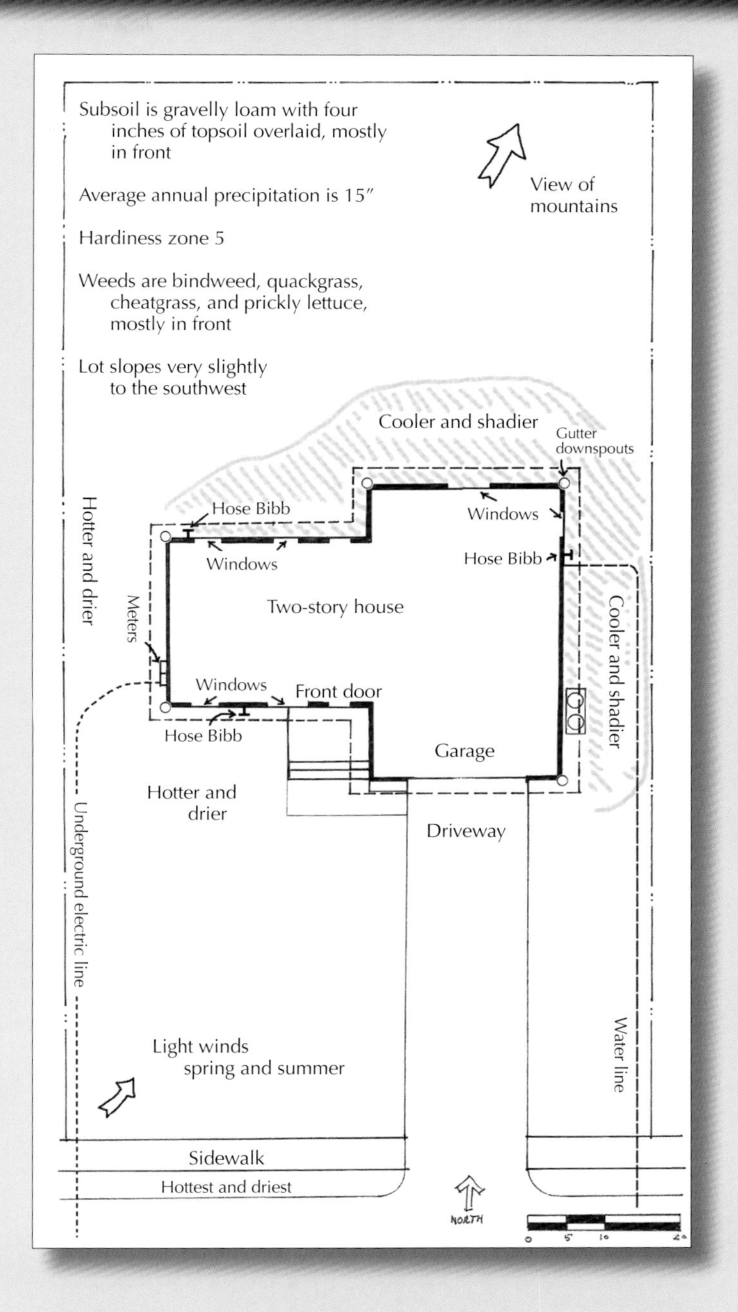

## How to Test Your Soil

*Texture*: Moisten a handful of soil and try to roll it into a cylinder. If it crumbles, the soil is sandy. If the cylinder forms but breaks when bent, the soil is loamy. If the cylinder can be molded into a curved shape, the soil is clay. Or shake a cup of screened soil with a tablespoon of powdered dishwashing detergent in a straight-sided bottle of water. Fine gravel will settle out in a few seconds, sand within 1–2 minutes, silt within 1–2 hours, and clay within 1–2 days. Measure the total thickness of soil in the jar, divide the thickness of each layer by the total thickness and multiply by 100 to get the percentage of each. A coarse soil will have >70% sand and gravel, while a clay soil will have >50% clay, and a silty soil will have >70% silt. Soils with <40% clay, <70% silt, and <60% sand are usually loamy. You can use a texture triangle graph (see resources) for a more accurate texture description.

*Drainage*: Drainage describes how quickly water moves down through the soil. Determine your drainage by digging a hole 16″ deep and filling it with water. If the hole drains within one hour, drainage is "rapid;" if the hole drains in a few hours, the drainage is "good;" if water stands for a day or more, drainage is "slow."

*Soil Chemical Properties and Fertility:* Inexpensive home soil test kits can give you a rough idea of the pH (acidity/alkalinity) of your soil and ballpark levels of major nutrients such as nitrogen, phosphorus, and potassium. These kits

of the landscaping. If your house is newly constructed, the landscape may present an essentially blank canvas that you can design from scratch. More commonly, there is already existing landscaping, if only the lawn and foundation shrubs installed by the developer. Older properties often have well-developed landscaping, including mature trees. These too can effectively be converted to native plant landscaping, but there will be many decisions to make. You will need to identify which plants you want to remove or keep, based on several criteria. For example, perhaps you currently have high-water-use plants in areas where low-water-use plants would be more appropriate. The existing vegetation may include invasive introduced plants, such as Russian olive or tamarisk, that would be best removed. And, if you plan to have an area dedicated to an exclusively native plant landscape, it will of course be necessary to remove existing usually involve mixing a small amount of soil with water, then adding a test reagent to the mixture and comparing the resulting color with a set of standard colors printed on paper. Test kits are sold in most places where you can buy gardening supplies, and are also readily available for purchase on the Internet. The results will not be highly accurate, but as you just want to find out whether your soil falls within broad ranges of acceptability for native plants, they will probably be adequate.

A pH range of 6.5 to 8.5 is suitable for most natives, with desert and semi-desert plants more tolerant of the higher pH. If the fertility level, especially the nitrogen level, falls in the "high" range for crop plants, it is probably too high for desert and semi-desert plants to perform well, though it should be all right for foothill and mountain plants. Don't worry if the nitrogen level is in the "low" range unless you are growing foothill and especially mountain plants. In this case, add weed-free organic amendments to the soil. Never fertilize native plants with chemical fertilizers. If your pH is very high (greater than 8.5), it is possible that your soil is sodic (too high in sodium)—this is a pretty serious problem for just about any plant. A white powder on the soil surface is also usually a bad sign, indicating possible high salinity. If you suspect that your soil is saline or sodic, it is best to get it laboratory tested (see resource list). These laboratories can also perform more rigorous versions of other tests.

non-native plants, including traditional turf, from that area. Make notes on the base map of any plants you plan to keep, as well as the extent and type of shade they are likely to create.

*Weed Problems*: You need to mark areas of weed infestation on the base map and indicate the identity of the problem weed. Weed problems can be a major obstacle to a successful native plant landscape, so it is important to identify the problem early and deal with it, preferably before the new landscape is installed.

Annuals like pigweed and cheatgrass can generally be controlled by hand pulling, though some species have seeds that persist for many years. Mulches can be used in the installed landscape to control annual weeds, and once the natives are well-established, they can suppress weed growth. Perennial weeds, particularly those with underground rootstocks or rhizomes, present more of a problem, as

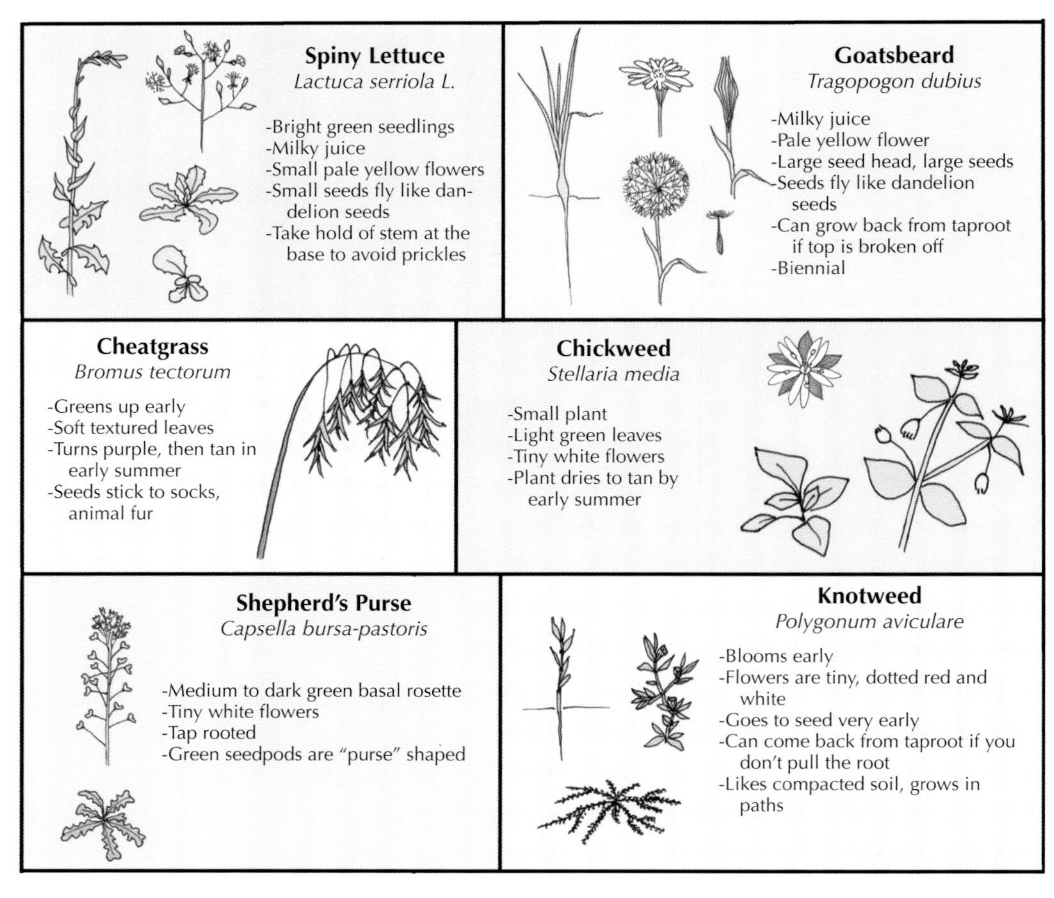


A few common spring annual weeds of the Intermountain West.

hand pulling is temporarily effective at best and mulches are not good deterrents. These problem perennials include field bindweed (wild morning glory), whitetop, quackgrass, and, yes, Kentucky bluegrass, as well as others, such as Russian knapweed and Canada thistle, that may be locally important. It is essential to do your best job possible of eliminating these weeds prior to installation. If you are removing a Kentucky bluegrass lawn, be sure not to leave behind living pieces of rhizome. To make matters worse, some of these weeds, such as field bindweed, also form persistent seed banks, so that vigilance is necessary to prevent reinfestation from seed.

*Microhabitats*: Patterns of microclimate variation are created as the effects of sunlight, temperature, snow cover, and wind are modified by topography, structures, and existing vegetation. These effects can have a major impact on the growing environment for plants. If this variation in microclimate is used wisely, the site can support a much broader array of native plants, and even different plant communities, than a site that is perfectly uniform. Areas of the property where the microclimate is expected to be exceptionally cool and shady, such as underneath a tree, or hot and dry, for example, a south or west

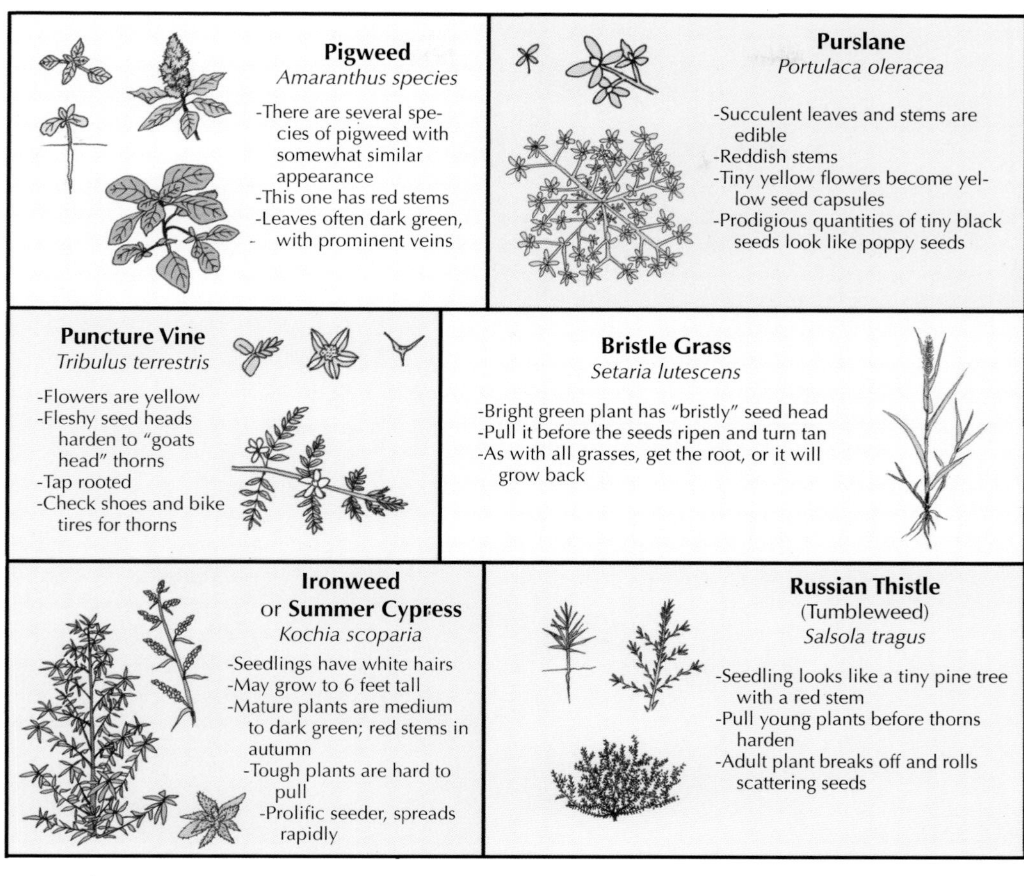


A few common summer annual garden weeds of the Intermountain West.

exposure, should be indicated on the base map. Remember to note the effects of buildings, fences, trees, and pavement, both those that are located on the site itself and those on adjacent properties. Variation in slope, aspect, drainage, and soils will also impact the microhabitats for plants and should be taken into account.

## Considering How You Will Use Your Landscape

Once you have looked at the characteristics of your site, the next step is to think about the human requirements for the design. Consider what kind of activities will take place on your property. Do you need a place for children to play? A vegetable garden? A dog run? A place to play basketball? Don't forget a place for mundane items like garbage cans, composters, or recycling bins. Another common outdoor activity to think about is entertaining, whether it is just the family around a picnic table or a full-fledged dinner party on the deck. A place to hang a hammock for reading in the shade could be another consideration, or a strategically placed bird feeder that can readily be seen from a window.

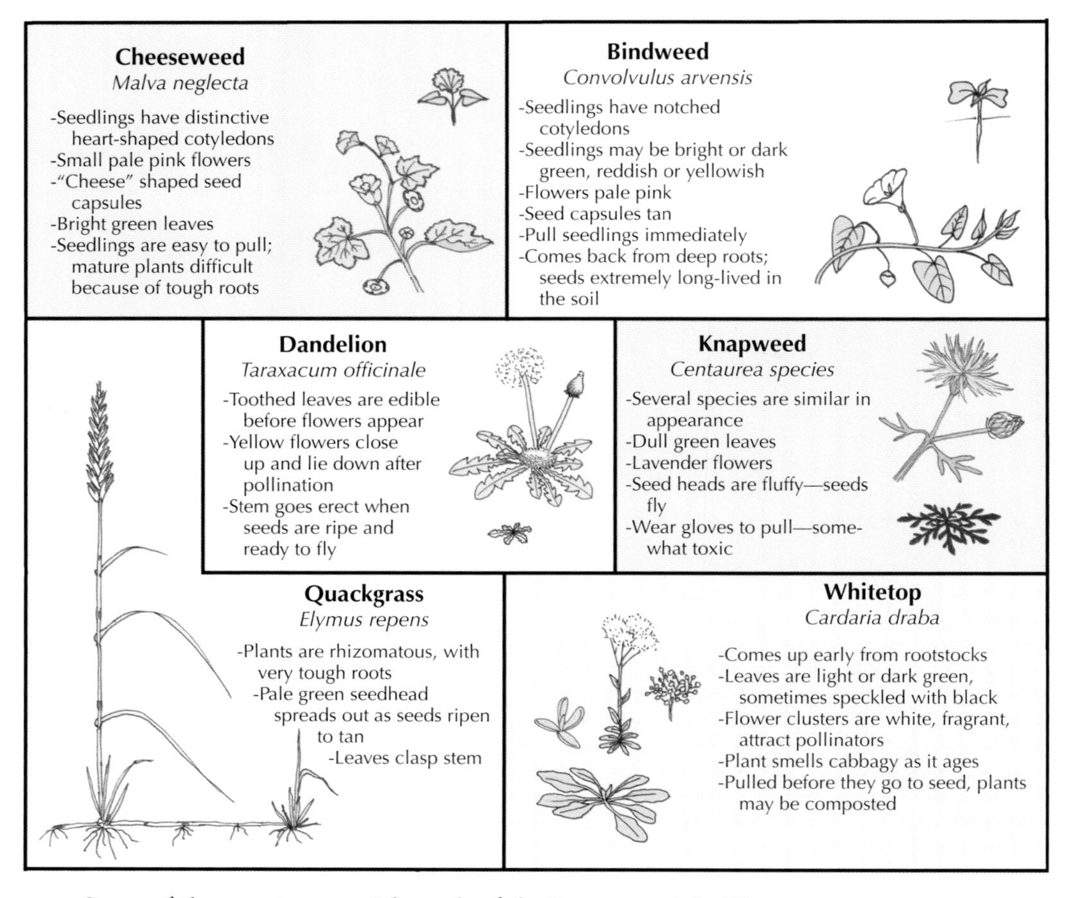


Some of the worst perennial weeds of the Intermountain West.

The landscape you are about to create will probably be much more interesting and inviting than those you have experienced before, so think about making it easy to access and enjoy. What is more appealing than the perfect nook that beckons from a far corner of the yard? A native landscape is a place to experience, rather than just a place to look at from the street or the front porch. You will want to go out into your landscape, as you would go to a wild place, just for the pleasure of being there. Take this possibility into account when considering how you will use your landscape.

You also need to think explicitly about how you want be able to move through your landscape. Identify destinations and consider the best way to travel from one to another. A large block of landscaping that offers no obvious access will not feel welcoming. At the very least, consider maintenance access for these areas.

The different areas around the front, back, and sides of a building often lend themselves to different uses. Even though you want to create a native landscape, there can still be a place for non-native plants that you want to keep. You can continue to have those daffodils you got from your grandmother. And there may still

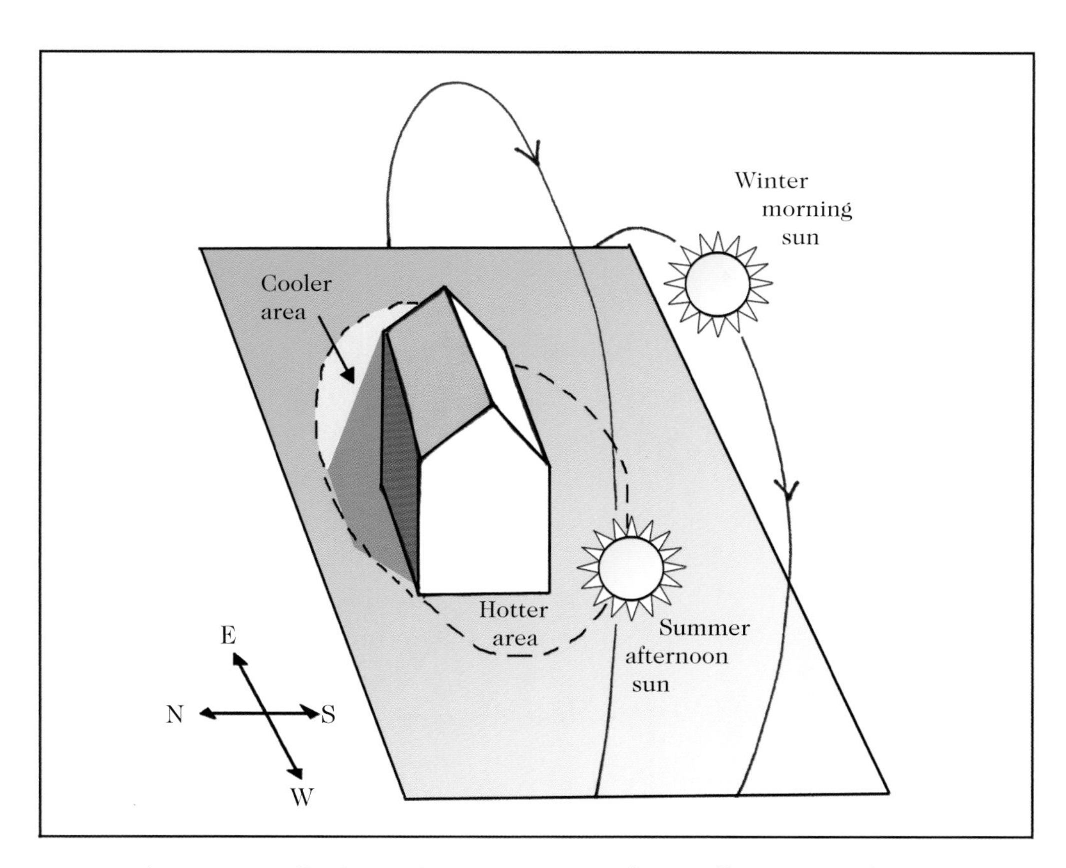


Your house is usually the single most important factor affecting microhabitats on your property, especially for new landscapes without mature trees. Because of the sun's path through the southern sky in winter, the south side of the house will be much warmer than the north side. Even in summer, when the sun is higher, the north side remains in shade for much of the day. The west side will be hotter than the east side, because it receives full sun during the hottest part of the day. You can plan to exploit these microhabitat differences by planting appropriate species, rather than trying to compensate for the differences with extra water. On the other hand, trees planted on the west side for shade will require water but will repay you by cooling the house.

be a place for lawn—it just needs to be restricted to the areas where it actually fulfills a function. It is even possible to successfully mix natives and non-natives in the same section of the landscape, as long as the species planted together have the same water, soil, and light requirements. But we strongly recommend dedicating whole areas of the site to exclusively native landscaping, in order to create the powerful sense of place that is the concept at the heart of this book.

Another factor to think about at this stage of the design process is how much time and money you want to put into your landscape. If you love to be outside tending to the plants, you will have a different approach than if you want simplicity and ease of maintenance. If cost is a limiting factor, you may want to plan to install your landscaping in phases. This also has the advantage of providing a learning experience whose early lessons can be applied in the later phases.

Now that you have walked through your landscape one more time and thought through the human activities that will take place there, it is time to get a fresh copy of the base map and sketch in these use zones and the circulation corridors (paths) that will connect them. This does not have to be a careful work of art. In fact, these kinds of sketches are called "bubble diagrams" for a reason.

## Planning Landscape Water Zones

Once you have a clear idea of the possibilities presented by your site, as well as a vision of how you want to use it, there is one more important step before you begin the actual process of designing your native landscape. That step involves making a provisional plan of how your landscape will be laid out spatially in terms of watering zones. As we discussed before, the most basic attribute that describes your site is its precipitation zone, but this may be modified locally by the effects of slope and aspect. The best clues about local climate are often provided by the plants themselves. The native plant community adjacent to your site, or on similar sites in your area, is probably the best indicator of the macroclimate you will experience. The plan for watering zones that you develop needs to incorporate this basic climate information, and it should also take into account existing microhabitat differences on the site. The plan can also call for modifying the site, either by using irrigation or water-harvesting to create habitat for plants from wetter places, or by modifying topography (slope and drainage) or soil to accommodate plants from drier places.

The most straightforward way to design a native plant landscape is to use plant communities characteristic of your precipitation zone as the inspiration for your landscape. This minimizes the need for irrigation or other habitat modifications, but limits the range of plant communities that you can use. If your site is located in the desert precipitation zone, for example, it will be easy for you to create designed landscapes that use plants from this zone to capture elements of the desert aesthetic, whereas creating foothill plant communities will require considerable modification, and creating mountain communities may be next to impossible. Conversely, if you live in the mountain precipitation zone, you will have an easy time creating a designed landscape that reflects mountain native plant community aesthetics, but it will be difficult or perhaps even impossible to create landscape conditions for true desert plants. If you live in the semi-desert precipitation zone, you will have more flexibility in creating a variety of native plant community aesthetic effects. In general, though, it will be easiest to establish and maintain a native plant landscape if the plan for most of the area is based on a suite of plants that would naturally be found in your precipitation zone.

A fundamental principle of native plant landscaping, and of waterwise landscaping in general, is to group plants with similar water requirements. Our system

## Considering How You Will Use Your Landscape

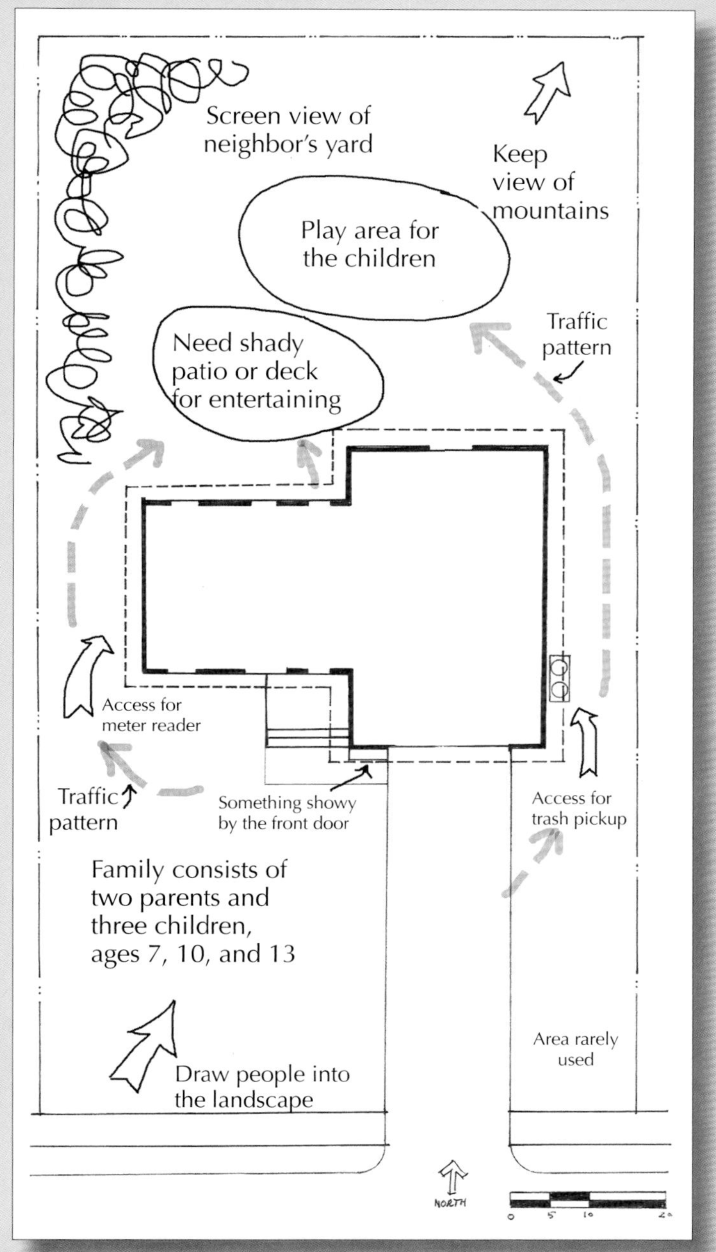


Think about the needs of the people who live in your house, and the kinds of things they like to do outside in the yard. Ask yourself and your fellow residents the following kinds of questions:

- Do you need a place to entertain outside? How formal would you like this to be?
- Do you need a place for small children to play? Does the back or front yard need to be fenced?
- Do you need an area dedicated to pets?
- Do you want to have a vegetable garden?
- Do you want to retain a space for traditional high-water-use flowers?
- Are there areas that will receive little use because of their location?
- What are the natural patterns of traffic flow through the landscape?
- How will the utility areas be accessed?
- Do you like to relax by working in the garden, or are you looking for super-low maintenance that suits your super-busy lifestyle?
- Are there particularly nice views, either from outdoor vantage points or from certain windows, that you want to preserve?
- Are there objects, either in your yard or in adjacent yards, that you would like to screen from view?
- How much is privacy in the back yard an issue?
- How important is shade as a way to cool your house in summer?
- How important is water conservation as a goal in your design?

for describing plant water requirements has five categories: minimal water, low water, medium water, high water, and very high water. The first four categories correspond to the precipitation zones described earlier for desert, semi-desert, foothill, and mountain plant communities. We define plant water requirements in terms of ranges of natural precipitation. If the species grows naturally in a desert environment where the average annual precipitation is ten inches or less, that plant has *minimal* water needs. If the species grows naturally in a semi-desert environment, where the average annual precipitation is between ten and fifteen inches, the plant has *low* water needs. Plants with *medium* water needs grow in the foothill zone, where annual precipitation averages fifteen to twenty inches, while plants with *high* water needs grow in the mountain zone, where the average annual precipitation is over twenty inches. Streamside and wetland plants from all precipitation zones could be described as having *very high* water requirements, as they generally need a surface soil environment that is wet much of the time regardless of precipitation.

To put these precipitation ranges in perspective, it helps to know that most cities and towns in the Intermountain West are located at valley edges in the semi-desert precipitation zone, where the average annual precipitation is ten to fifteen inches. Communities on the benches and in the high valleys are mostly in the foothill zone, with fifteen to twenty inches of annual precipitation on average. The high precipitation zone is found up in the mountains, but even there precipitation rarely averages more than forty inches per year. It is educational to compare these figures with yearly averages for some cities in other parts of the United States and Europe—cities like New York, Chicago, Washington DC, and Seattle average between thirty and forty inches, putting them well into the high precipitation zone in our system. Virtually all the major cities of northern Europe also experience what we define as the high precipitation regime. At the other extreme, warm desert cities like Las Vegas (five inches) and Phoenix (eight inches) are well below the threshold for the desert zone, and even Los Angeles barely makes it into the semi-desert zone with twelve inches per year on average.

The take-home message here is that in the mountain precipitation zone, many traditional garden plants can grow with little supplemental water, as they do in the eastern United States and in northern Europe, but when these same plants are grown in a semi-desert or even a foothill water zone, substantial water must be added in the form of irrigation to make up the difference. On the other hand, if we use plants that can prosper in nature in the desert zone, it stands to reason that they will rarely if ever need supplemental water after establishment if planted in a location within the semi-desert zone. The total water provided to the plant is essentially the sum of natural precipitation and supplemental irrigation (though harvested water and subsurface ground water are also part of the

## Sketching in the Landscape Water Zones

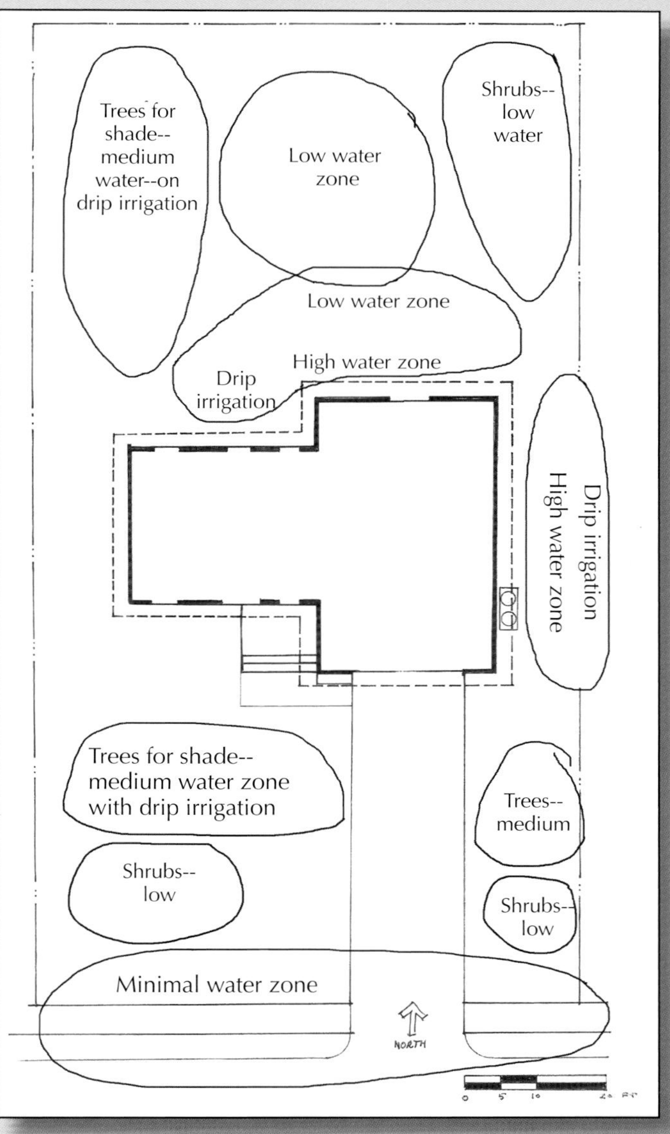


Your preliminary ideas about water zones should be based on the characteristics of your site, especially the precipitation zone where it is located, and on the expectations you have for your landscape. This example is located in the semi-desert precipitation zone with 15″ of annual precipitation.

- The parking strip and the area next to the front sidewalk receive little use and are a long way from a hose bibb. Making these areas a minimal water zone means that they will virtually never need supplemental water. Because your site is sloping and well-drained, you will not need to make special provisions to grow minimal-water-use plants.
- The shadier, cooler areas on the north and east sides of the house are a logical place for a high water zone. This also places a more lush planting adjacent to the back door, a high-use area. This planting grades into a low-water planting to the north.
- Trees are useful as a screen in the northwest corner of the lot and will also serve to shade the high-use area near the back door. Trees to the south and west of the house will help keep it cool in summer. Using medium-water-use trees from the foothill zone means you can achieve this with less added water.
- The rest of the property will be landscaped with low-water-use plants that will rarely need supplemental water. Some of these plantings are tentatively shown as shrub plantings.
- All the blank areas on the plan are assumed to be in the low water zone, which corresponds to the precipitation zone at the site.

total available water in some situations). If post-establishment needs for a plant are met by natural precipitation, there is generally no need to add supplemental water. This principle forms the basis for the watering recommendations in this book. The idea is to "top up" the water that is provided as natural precipitation with enough inches of supplemental water to reach a total that approximates the water requirement of the plant.

In general we recommend "topping up" to the maximum for the precipitation zone of origin, in the interests of keeping the plants looking their best. If you are located in the desert zone, for example, and want to grow desert plants, we recommend that you "top up" the natural precipitation to the level of ten inches. The translation is that, after establishment, these plants would rarely need supplemental water and then only under drier-than-average (drought) conditions. If your site is in the semi-desert zone, plants from the desert zone, designated as minimal-water-use, will virtually never need supplemental water after establishment, while those designated as low water use (semi-desert zone) may benefit from occasional watering, up to five inches of added water per growing season, to "top up" the total water to fifteen inches. Plants designated as medium water use will need regular but infrequent watering (five to ten additional inches in an average growing season), while those designated as high water use will need to be provided with ten to fifteen inches of supplemental water per growing season. Compare these water requirements to those for a Kentucky bluegrass lawn in the semi-desert zone, which requires twenty to thirty inches of supplemental water a season to maintain an acceptable summer appearance.

A good thing to remember is that precipitation varies dramatically from year to year in any given spot, and wild plants for the most part have little trouble coping with this variation. This means that you just have to get ballpark close in your watering zone planning, not accurate down to the last inch. You need to focus on three things. First, native plants vary widely in their water needs, depending on what precipitation zone they come from. Second, the water that falls out of the sky as natural precipitation does count toward meeting those water needs. You only have to make up the difference. And third, there is definitely such a thing as too much water.

Overwatering can be at least as much of a problem in a native plant landscape as underwatering. When moving desert and semi-desert plants into precipitation zones with more water, it is often necessary to modify site conditions to compensate for the excess water that these plants will be receiving naturally. Placing desert and semi-desert plants in the hottest, sunniest, driest microclimates, adding inorganic amendments to make a coarse soil texture or replacing the soil entirely with a sand-gravel mix, making sure the soil does not contain excess organic matter, and placing the plants on berms or slopes to improve drainage

## Plant Water Needs and Precipitation Zones

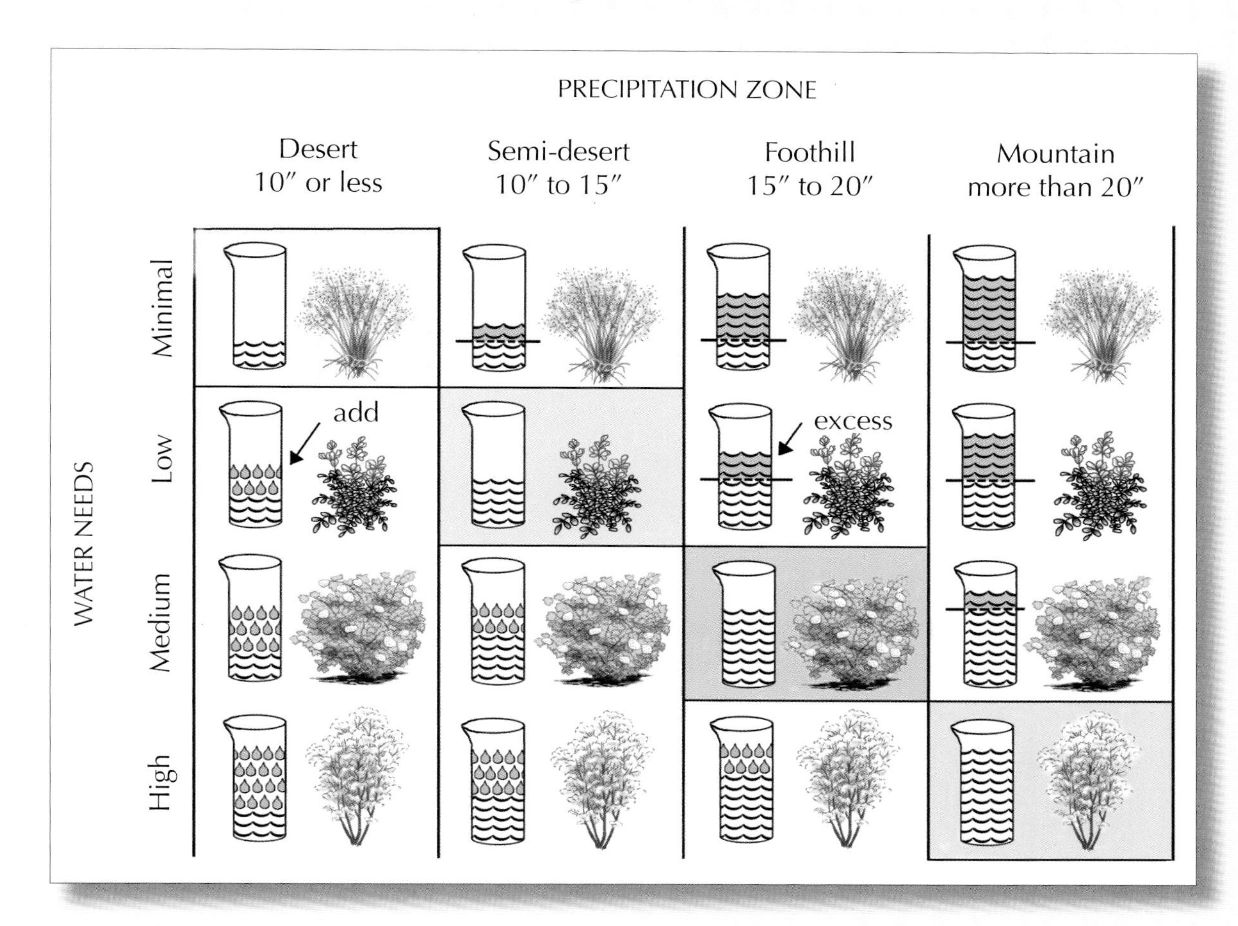


The water needs of a plant in your landscape depend on the relationship between the precipitation zone where it grows in nature and the precipitation zone at your site.

- If these match, the plant will rarely need extra water, and it will not suffer the negative effects of too much water. Plants along the diagonal in this diagram are matched with their site precipitation zone.
- If the plant is from a zone wetter than the zone at your site, you will need to add water to meet its needs (blue added water in beakers below the diagonal in this diagram).
- If the plant is from a zone drier than the zone at your site, it will on average be receiving more water than it needs, just from natural precipitation (pink excess water in beakers above the diagonal). Not only will it need no added water, it may also need to be protected from the effects of too much water.

are some tactics that improve their chances for survival. It is often quite difficult to grow plants in environments more than two precipitation zones removed from their optimum zone. This means that growing typical desert plants in mountain environments or typical mountain plants in desert environments should generally be attempted only by experienced gardeners, if at all. It also explains why it is nearly impossible to grow desert and semi-desert natives in the context of a very-high-water-use traditional landscape.

As you plan your watering zones, bear in mind that even native trees generally have medium or high water requirements. Many people find a landscape without shade uninviting at best, so that high water zones with a few well-placed trees are often a good idea even in a semi-desert or desert landscape. But trees are not the only way to create shade. Taking advantage of shade created by existing buildings, or including shade structures such as trellises, awnings, ramadas, or arbors, are some other design options.

Now it is time to sketch your ideas for the watering zone plan onto another copy of the base map. This means designating each area in the landscape as a minimal, low, medium, high, or very high water zone. Very high watering zones include those that will be maintained in traditional lawn, traditional annuals and perennials, and vegetables. The remaining zones correspond to the precipitation zones described earlier. You will need to plan for supplemental water to support plants in water zones that need more water than is provided by the natural precipitation on your site. And you will also need to plan to counter the effects of excess water for the plants in water zones that need to be much drier than your precipitation zone.

The watering zone plan is not cast in stone at this point. The reason for considering water first, before carrying out the aesthetic activity of creating the design, is to make sure that you take water into consideration at the most fundamental level. Most landscape designers have worked in environments that are not water-limited, whether because of adequate natural precipitation or because of a cultural perception that unlimited supplemental water is available. This is why design dogma has always held that landscape choices can be determined primarily or entirely by aesthetic considerations. But if you want your intermountain landscape to truly be 'of the place', it will need to reflect the way that plant community patterns are shaped by water in nature. That means carefully considering how the designed landscape will also be shaped by water.

## Creating the Mass/Space Diagram

The next steps in designing a native landscape require a degree of melding. After mapping existing site conditions on the base map, considering how you will use your landscape, and thinking about watering zones, you are ready to develop a mass/space diagram. In a mass/space diagram, *spaces* are the open areas: pathways, decks, terraces, dry stream beds, and areas of low vegetation or mulch. *Masses* are areas with vegetation, relief, or structures that rise above the spaces. These may include shrubs, tall grasses or perennials, large rocks, raised planting areas, and canopy cover. Canopy cover consists of masses that rise well above eye level. They may be formed by trees or may be constructed, such as trellises and arbors. It is also possible to conceive of mass and space in three tiers: open areas or space, mid-height masses, and canopy masses. Try to visualize the area in three dimensions, and hearken back to the wild landscapes that formed the heart of your original design conception. What were the mass/space relationships of those wild landscapes, and how can you re-create that feeling of mass and space in your designed landscape? Remember the idea of landscape patterns on multiple scales. How can you recreate the ambience of a grand landscape using similar mass/space relationships, only on a smaller scale?

You are now ready to tap into your "inner designer," and, trust us, you have one in there. At this stage of the process, it is important to be as "fluid" as possible, allowing spaces to move river-like through the composition, and masses to drift in patterns that reinforce the flow of space. As you are diagramming these patterns, it can be helpful to listen to some "flowing" music as an inspiration. Make several quick mass/space designs in chalk pastels or colored pencils on thin-paper overlays over the base map, or work directly on photocopies, using different color tones for different kinds of masses and spaces. It can also be helpful to walk around in the area and physically feel the flow of the space. Now is the time to remember mystery, complexity, coherence and legibility. And stay loose—no one is going to judge the results of this exercise except you.

The mass/space diagram is not created in a vacuum. Instead, it represents a distillation of all the information you have gathered and an integration of this information into the design of an aesthetically satisfying landscape. Give the process of creating the mass/space diagram enough time. If you find this design exercise difficult, just think of your first few efforts as practice—you will get the feel of the process.

When you have completed a number of provisional mass/space studies, you are in a position to select one of the studies as "the best," or you may incorporate features from several of them to blend together in a more refined mass/space diagram. The importance of the mass/space diagram cannot be overstated. It becomes the basis for the remaining steps in the process of designing a landscape. It can make the difference, ultimately, between a unified, harmonious overall landscape composition and a patchwork collection of plants, rocks, and other landscape features. It is this process of creating a unified composition that will transform your landscape into a place, one that will invite you to enter and encourage you to stay and explore.

## Laying Out the Landscape Plan

The mass/space diagram you have created provides a focal point for the next steps in the process of designing your landscape. Now you need to think more concretely about how these mass/space relationships will be translated into paths, use areas, landscape features, and planting areas on the ground. You have already considered many of these issues, and probably have a good idea of what most of the masses and spaces on your mass/space plan actually represent. For example, you have thought about how you will move through the landscape. The mass/space diagram helps you lay out these pathways in a manner that is aesthetically appealing as well as functional. You have considered where you need screening and where you need to leave openings for appreciation of more distant views. The mass/space diagram helps you visualize how you will incorporate these screens and openings into a harmonious design. You have decided which areas you will dedicate to native plant landscaping as well as determined whether some areas will be maintained in more traditional landscaping. The mass/space diagram helps you make sure that contrasting areas are connected and integrated into the overall design, rather than representing discordant elements. And you have thought about the kind of place you want your native plant landscape to be, a place that is experientially rich as well as ecologically sound. The mass/space diagram helps you see how you will achieve that goal, by echoing the mass/space relationships that you have found to be beautiful in wild landscapes, and by providing a way for you to enter the landscape and be comfortable and at home there.

You have also thought at least provisionally about site modifications you might need to carry out in order to accommodate plant communities from wetter or drier precipitation zones. This may include providing supplemental water in zones for higher-water-use plants, as well as ameliorating the effects of too much water in zones for lower-water-use plants. Water zones are often implicit in the configuration of mass and space; now is the time to make them explicit. Mid-height masses may be areas of shrubs, tall grasses, or perennials; they may represent rock groupings or hardscape structures; or they may be areas that are planted with low vegetation but that are topographically raised, either naturally or through the creation of berms or retaining walls. As mentioned earlier, canopy masses may represent trees or shade structures, such as gazebos, arbors or trellises. And space may be created as paths and other hard surfaces, as areas of ground cover or other low vegetation, or as areas of mulched open ground, as in the dry wash concept that many find appealing. Remember that canopy, mid-height mass, and space can be somewhat relative

## Creating the Mass/Space Diagram

This is the first step of the design process where aesthetic considerations become the most important element.

- The goal is to create a visually pleasing design that integrates masses and spaces into a configuration that exhibits organic unity.
- Mass/space diagrams often use color to distinguish between masses and spaces.
- Tentative shapes for the canopy masses can be roughed in if desired.
- Make several quick sketches, then combine and refine them until you have a finished pattern that pleases you.

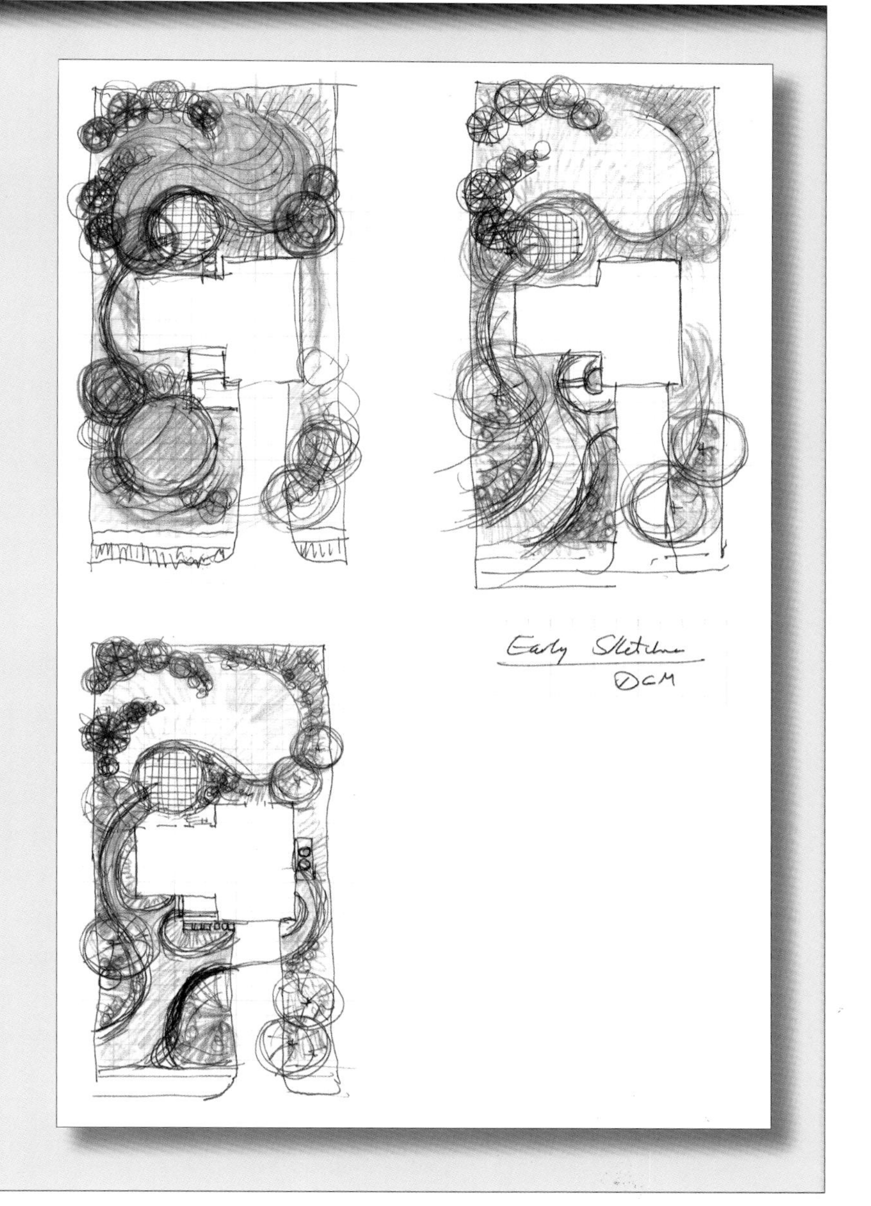


concepts. In a forest of tall trees, shrubs eight feet tall could function as mid-height mass, but in a desert landscape, an eight-foot-tall tree or arbor would definitely function as canopy.

Which option you select to represent space, mid-height mass or canopy for a particular area will depend partly on how you plan to use the area. It will also depend on the water zone and the native plant community that provides inspiration for the area. Water zone and native plant community are, of course, inextricably linked, and together they provide the underlying logic for the design of the planted areas.

Your next step in the design process is to take another copy of the base map and lay out your landscape plan. Basically, this involves recording the decisions you have made about the space, mid-height mass, and canopy mass relationships in the mass/space diagram, that is, how you will translate each area of space, mid-height mass, and canopy mass into paths, use areas, landscape features, and planting areas. To do this properly, you will need to take into account all the planning you have already carried out, as recorded on earlier versions of the base map—namely your site inventory, bubble diagram of human use patterns, and water zone map—as well as your most refined version of the mass/space diagram. Spread all these out on the table, and keep your creative juices flowing, because this is the step where you will decide how your landscape will actually look and feel. First, lay out all the areas that are or will be hardscape—the paths, patios, and decks, as well as any other paved or finished surfaces. In areas where topographic modification is planned, show the project as completed, for example, a berm or raised bed or a terraced area behind a retaining wall.

Next, consider the non-hardscape areas on your mass/space diagram. In areas that are designated as canopy, which will be planted to trees and which will have constructed canopy, such as trellises or arbors? Check on your water zone map to see how you had intended to water a particular canopy area. If it is in a high-water-use zone, you will be able to plant trees from the foothills or mountains, whereas if it is in a medium-water-use zone, you will be limited to trees from the foothill woodland or mountain brush communities. If the canopy area is designated as a low or minimal water zone, you will need to use constructed canopy. Of course, it is possible to change water zone designations to fit your revised concept of mass/space and plant community configuration. Remember, the water zone map was intended to be provisional. But also remember to take microhabitat considerations into account. There may have been a good reason to designate the area shown as canopy as a low- or minimal-water-use zone. You may need to check your site inventory to jog your memory on this. If the area is not suitable for trees but definitely needs to be canopy in the design, constructed canopy can be an attractive and functional solution.

In non-hardscape areas that are designated as space, your decision on how to create this space should again take watering zones into account. These areas may be planted, or they may be mulched bare ground. If the area is designated as very high water, it can support traditional lawn as a ground cover. In fact, it may already support lawn that you intend to keep, which would be one of the few reasons for designating a very high water use zone on the water zone map. Spaces in medium- or high-water-use zones can support low ground covers

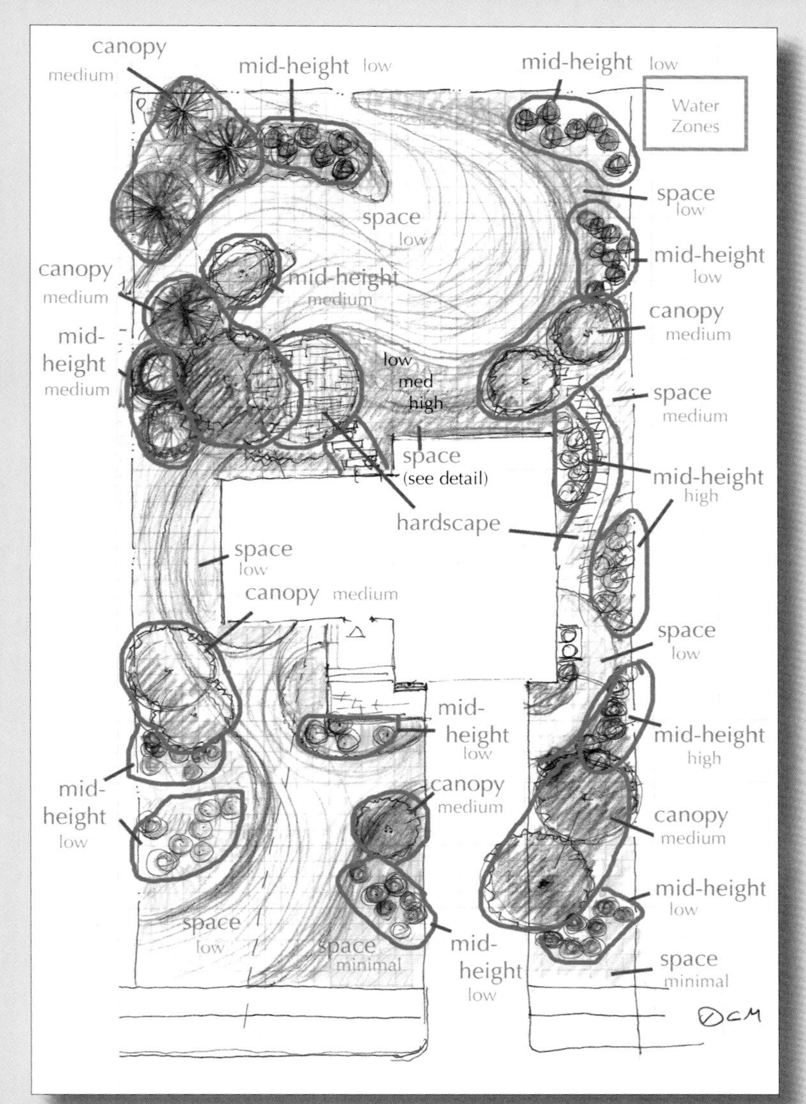


## Laying Out the Landscape Plan

The landscape plan makes the masses and spaces in your mass/space diagram into a concrete design, with hardscape features, mulched open areas, and planting areas representing canopy, mid-height mass and space. Each planting area is designated as belonging to a water zone. Notice that the placement of the individual plants inside each planting area is still somewhat tentative. Once you select the species and determine the number of plants of each, a formal planting plan can be prepared.

- This design possesses minimal hardscape other than the driveway and front sidewalk.
- Decorative stone paths lead from the east-side utility area to the back yard and from the back door to the circular stone patio. Remaining pathways are planted with low vegetation.
- Areas of canopy and mid-height mass surround a large play space in the back yard, while the back door area features low perennials for a cozy but open feeling.
- In the front yard, a sweeping planted pathway draws attention through a mixed planting of low-water-use perennials accented with shrub and canopy areas, and ties the front and back yards.
- The east side features low ground cover and mid-height shrub masses, with another canopy area to the south.
- Low perennials fill the spaces between the shrub and canopy plantings.

such as blue grama (which can be mowed as a lawn grass), trailing daisy, or mat penstemon, as well as shorter shrubs, grasses, and perennials from foothill and mountain communities. But much of the space in the design will likely be designated as low or minimal water use, because the small plants of desert and semi-desert communities best lend themselves to creating the impression of space.

If you live in the foothills or mountains and do not want to use water-loving ground covers to create space, your best option may be to use mulched bare ground. Make sure that the mulched areas look inviting, not hostile. The cobble fields that are becoming popular under the name "rock lawn" definitely do not qualify as inviting. A coarse organic mulch, such as bark chips over a weed barrier fabric, would be a much better solution.

Last, consider the areas on the mass/space diagram that are designated as mid-height masses. For areas that will be planted to mid-height masses, the water zone may be low, medium, or high. Few if any minimal-water-use plants are tall enough to serve as mid-height mass, but some of the taller shrubs of the semi-desert zone would certainly qualify, and there are many large shrubs in the foothill and mountain zones. If you have or plan to create topographic relief on your site, you can sometimes use it to create the feeling of safe enclosure that would be achieved with a mid-height planting. Landscape features, such as rock groupings or structures of various kinds, can also be a part of mid-height mass.

Before you finalize your landscape design, check to make sure that you have incorporated all the relevant information generated during the process of inventorying your site, considering how your landscape will be used, deciding on site modifications, and sketching in watering zones. Now is the time to perform a quick quality control on the design process to this point. Your finished landscape plan should show all existing and planned hardscape, landscape features, and structures; each area of mulched bare ground; and each area to be planted. Each of the planting areas should be described as space, mid-height mass, or canopy, and each should have a water zone designation.

Now is the time to hearken back once again to the plant communities that provided inspiration for this whole process, and to designate each of the native planting areas in the landscape design as a particular plant community. This does not mean that you will create a perfect replica of any plant community. It just means a return full circle to the idea of taking your inspiration from natural landscapes. Even if you have limited experience with native landscapes in the Intermountain West, by following this design process and consulting the descriptions of natural landscapes provided earlier, you will be able to choose plant communities that are true to the aesthetics of your design as well as being adapted to the water zones and microhabitats your site can provide.

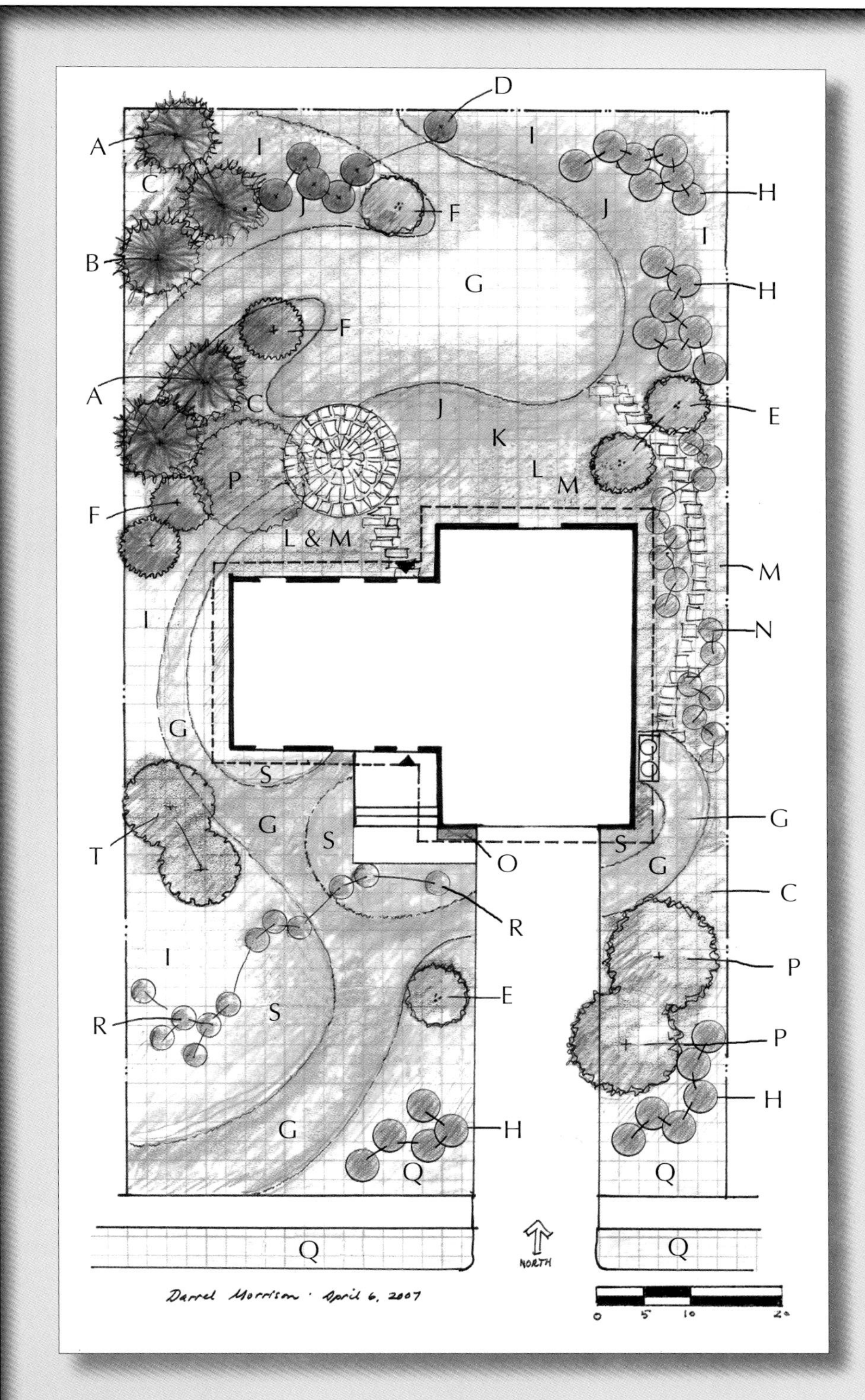


## The Final Planting Plan

A – Rocky Mountain Juniper
B – Pinyon Pine
C – Shining Muttongrass
D – Cliffrose
E – Utah Serviceberry
F – Alderleaf Mountain Mahogany
G – Blue Grama
H – Green Mormon Tea
I – Grass Mix:
Bluebunch Wheatgrass
and Alkali Sacaton Grass
J – Flower/Shrub Mix:
Hopi Blanketflower,
Firecracker Penstemon,
Bridges Penstemon,
and Winterfat
K – Transitional Flowers:
Utah Sweetvetch
and Lewis Flax
L – Mountain Flowers:
Sticky Geranium and
Mountain Puccoon
M – Creeping Oregon Grape
N – Mountain Snowberry
O – Desert Four O'Clock
P – Bigtooth Maple
Q – Desert Mix:
Indian Ricegrass,
Sundancer Daisy,
Silver Daisy, Showy
Sandwort, and
Shortstem Buckwheat
R – Rubber Rabbitbrush
S – Low Water Mix:
Utah Penstemon, Winterfat,
Palmer Penstemon, and
Hopi Blanketflower
T – Gambel Oak

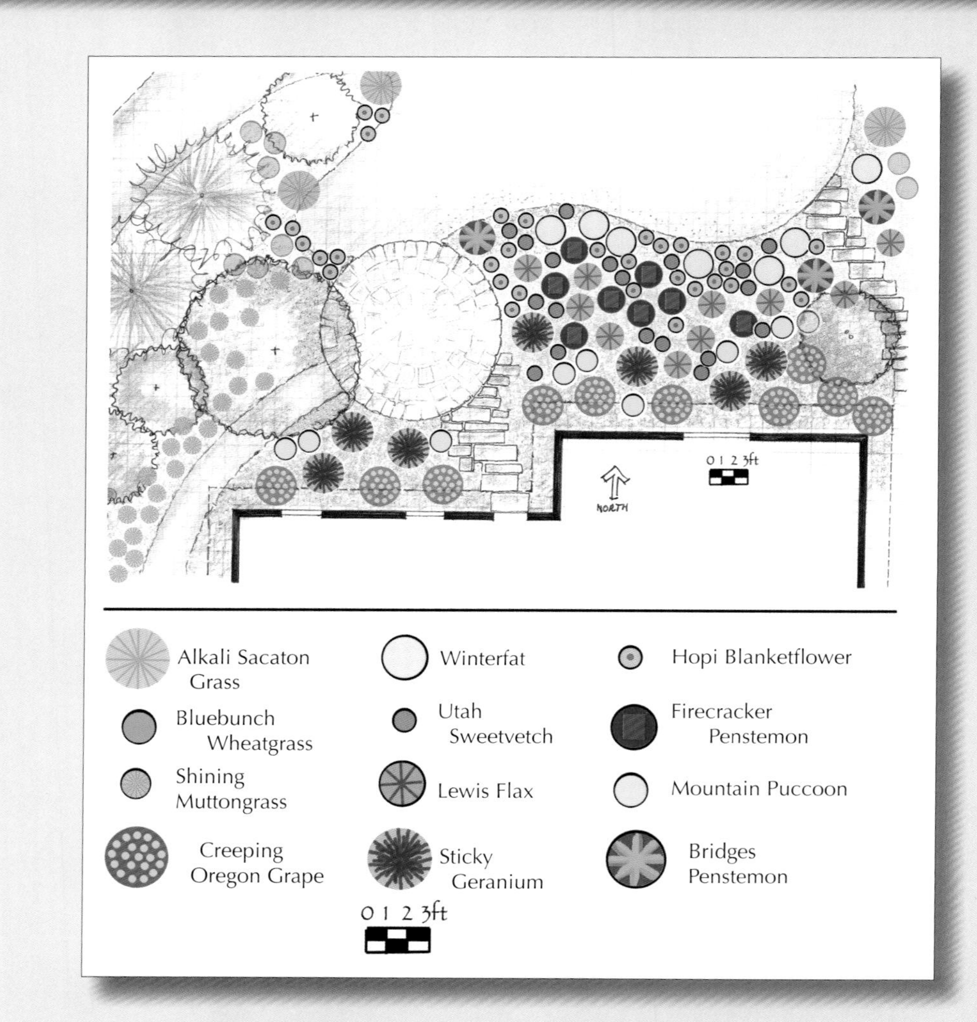


**Planting Plan Detail—Back Yard**

This plan shows the placement of individual perennial plants in the back yard planting.

- Notice how low-water-use perennials are used in the understory of the medium-water-use canopy planting.
- The perennial planting grades from high to low water and from shady to sunny as you move out from the north wall of the house. The middle section contains a transition buffer of low-water-use plants that can tolerate more water than those on the southern edge.
- Individuals of some species are arranged in drifts that overlap with the drifts of other species, creating a naturalistic meadow-like design.

## Choosing the Plants

If you are like most people, the first thing you did when you opened this book was to turn to the Plant Palette section and browse through the pictures of plants. Now it is time to revisit this section of the book with a more systematic goal in mind, namely, selecting the plants to populate the native plant communities that will occupy the different water zones in your landscape. Using your landscape plan as a basis, make a list of the different planting areas, along with information on mass/space relationships, water zones, and the approximate size for each area. It is usually easiest to think about the canopy areas first. For each area marked as canopy, consider the water zone that you intend to implement there.

### *Foothill Zone Plant Selection Example*

Suppose you have planned a canopy area that is in the medium water or foothill plant community zone. What is your vision of the plant community you would like to create there? Perhaps you love the sight of maple-cloaked foothills aflame in autumn, and have chosen a mountain brush community featuring bigtooth maple for one or more planting areas in your foothill water zone. Bigtooth maple represents both a dominant species and a "visual essence" species in the mountain brush community, and is an excellent choice for a canopy species in a medium water zone.

A planting composed of a single species, bigtooth maple, would meet the design requirements for canopy structure and coherence, but such a planting would definitely be lacking in complexity, with only one structural layer and only one type of foliage texture and color. In natural mountain brush communities, the maples are mixed with other species. Adding a few additional species will make the composition more interesting, especially if the plants chosen have contrasting forms, textures, and colors. You can find some of your options by going to the medium water section of the woody plant group in the Plant Palette, or by looking on the table provided for woody plants. Perhaps you want to add another tree, such as Gambel oak or netleaf hackberry, to the suite of species. These trees both have crooked trunks with rough bark that would contrast nicely with the straight, smooth trunks of the maples, and they also have leaves that turn gold in the fall, rather than red, as the maples do. Or perhaps you would like to add flowering shrubs like Utah serviceberry, mallowleaf ninebark, or littleleaf mockorange to the mix. These provide interest by blossoming in early summer, and they have a fine foliage texture that would contrast with the somewhat coarser texture of the maple leaves. They also present contrasting forms—the upright vase shape of the Utah serviceberry is quite different from the tight mounds of the ninebark and the arching, fountainlike shape of the littleleaf mockorange. These shrubs

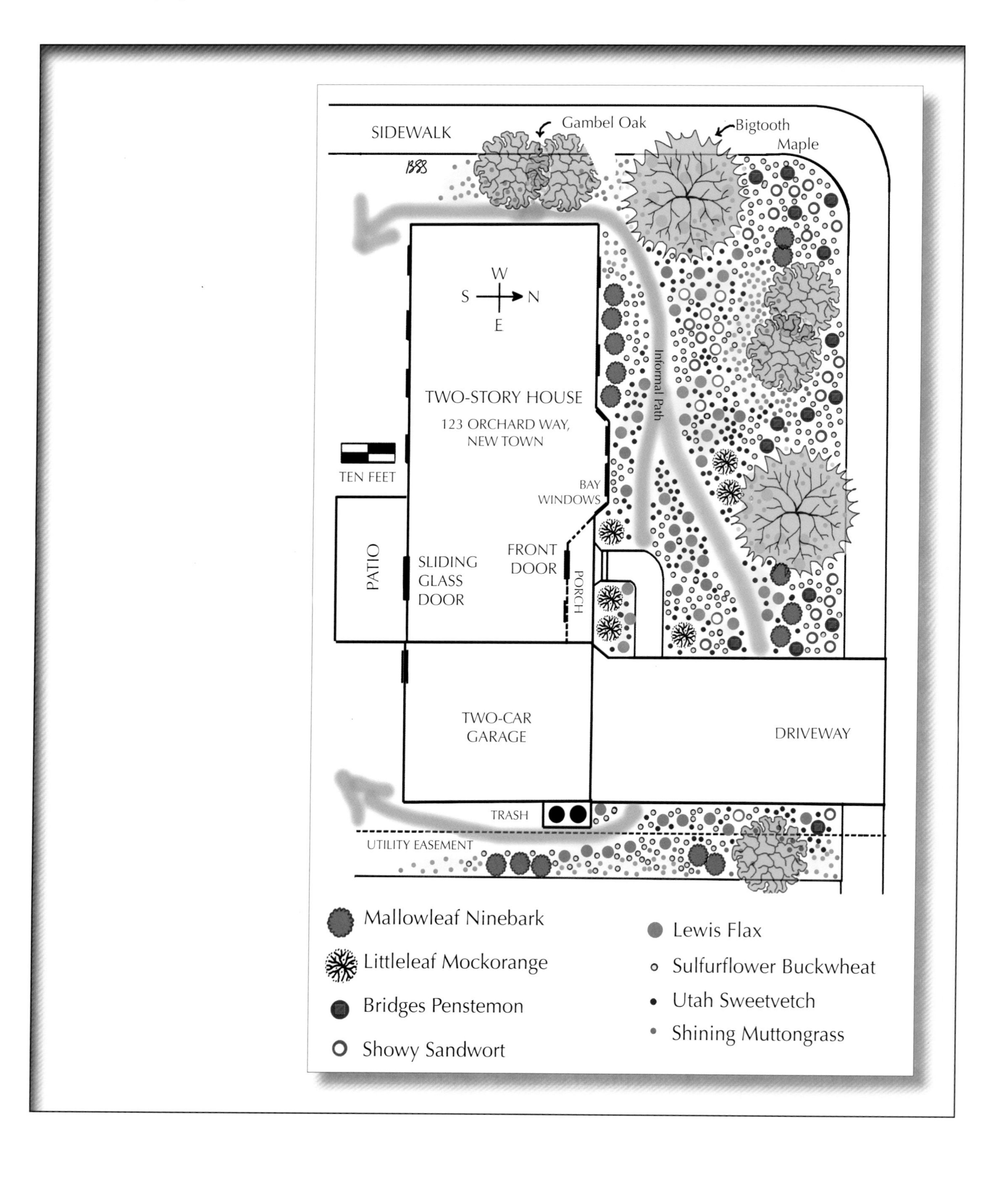

SIDEWALK
Gambel Oak
Bigtooth Maple
W
S
N
E
TWO-STORY HOUSE
123 ORCHARD WAY, NEW TOWN
Informal Path
TEN FEET
BAY WINDOWS
PATIO
SLIDING GLASS DOOR
FRONT DOOR
PORCH
TWO-CAR GARAGE
DRIVEWAY
TRASH
UTILITY EASEMENT
Mallowleaf Ninebark
Littleleaf Mockorange
Bridges Penstemon
Showy Sandwort
Lewis Flax
Sulfurflower Buckwheat
Utah Sweetvetch
Shining Muttongrass

can tolerate either full sun or the partial shade created by the canopy trees as they grow. The shrubs also grow more quickly, so they will provide structure and interest in the period before the trees attain their mature size.

Another consideration in the mountain brush community planting is how you will treat the ground beneath the trees and shrubs. One possibility is mulch, and in this case, something organic like bark mulch would be appropriate. As this mulch breaks down over the years, it will be replaced with natural mulch created through leaf fall from the planted species. Another possibility is to use shade-tolerant perennials in the understory. In nature, you will often see a combination of understory cover and mulched bare ground under the canopy and mid-layer woody species in a mountain brush community, and that is probably the best solution. If the areas surrounding the trees are kept free of competing vegetation, the trees will grow much faster. This is a good reason to keep the understory species in the interspaces. A good choice to represent an abundant understory species is shining muttongrass, which is shade tolerant, stays green all season, and is especially attractive when the pearly pink flowering stalks appear in very early spring. Possible wildflowers to plant in drifts among the grass clumps in the openings are dusty penstemon, Utah sweetvetch, sundrops, and Lewis flax. Some of these species are listed as low-water-use plants in the Plant Palette. But, as discussed under cultural requirements in the Plant Palette section, a species can usually be planted in the next higher water zone without a problem, for example, plants from the low-water-use category can be planted in the medium water zone. All of these wildflowers will reproduce readily from seed and will move around over the years into even more naturalistic arrangements. You could also use some understory species that are not shade tolerant, such as shortstem buckwheat or showy sandwort, to increase the initial complexity of the understory planting, with the knowledge that these would probably disappear over the years as canopy shade increases.

The total number of species to use depends on the size of the area—in general, larger areas can support higher diversity without looking cluttered. It is important to use enough individuals of each species to avoid the polka-dot penny candy look—one of these and one of those. You should end up with a list of perhaps seven to fifteen species, depending, again, on the size of the area and whether the understory will be planted. If you have several foothill water zone areas that need to be canopy in the mass/space diagram, for purposes of coherence it is usually best to use the same or nearly the same species assemblage in multiple areas.

### *Desert Zone Plant Selection Example*

Next, we will work through a plant selection example for a minimal water zone planting. Let's suppose that you love and live in the desert, and that you are intrigued by the possibility of creating a livable and attractive landscape using a minimum of added water. You have taken advantage of microhabitats, some existing and some created, to increase the palette of plant species you will be able to use in this landscape. By constructing a wash area that collects runoff from the roof of your house, you have provided habitat for some shrubs that can create mid-height mass. And by adding areas of landscape rocks artfully arranged on top of bermed earth, you have created additional mid-height mass and have also provided a water collection system of rocks to provide extra water for the species planted into this "rock outcrop" as well as at its base. You have three kinds of planting areas in this desert landscape—the open flat areas, the wash, and the rock outcrops. You also have an area with created canopy—an informal ramada (shade structure) that will let you enjoy the views of your landscape from a shady retreat. How do you go about selecting species for these different areas?

First look at species options for the open flat areas, which will provide the Zen-like feeling of space that is the essence of a desert garden. You will probably want to keep the composition for this part of the landscape quite simple: perhaps two or three species of small shrubs and a pretty clump-forming grass with two or three kinds of perennial wildflowers for the interspaces. Shrubs in the Plant Palette for the minimal water zone include shadscale, desert sage, big sagebrush (the Wyoming form is best for desert landscapes), winterfat, lacy buckwheatbrush, and sand sagebrush. All of these shrubs feature some version of the fine, silvery green foliage color typical of desert shrubs, but they differ in form and in other features. Desert sage is at its best in late spring, when the plants are covered with blue-purple blossoms, while the shell pink or sulfur yellow flowers of lacy buckwheatbrush appear in late summer to autumn, as do the satiny, rosy fruits of shadscale and the luminous white cotton wands of winterfat. If your site is sandy, then sand sagebrush is an obvious choice, as it is the quintessential "visual essence" species of desert sand dunes. Possible choices for perennial understory species in the desert landscape include Indian ricegrass, gooseberryleaf globemallow, sundancer daisy, Utah penstemon, and Indian paintbrush.

The rock outcrop area in your desert landscape gives you an opportunity to provide a focal point of diversity and interest. Suitable plants for the base of the outcrop include green Mormon tea, dwarf or datil yucca, and some of the taller perennials, such as prince's plume and Palmer penstemon. For the pocket planting areas that stud the rock outcrop, some choices of plants with compact growth forms but contrasting foliage would be showy sandwort, silver daisy, and silver buckwheat.

For the wash areas of the desert landscape, you need to pick large shrubs that will really add some volume to the mid-height masses. Good choices would be Apache plume, rubber rabbitbrush, and oakleaf sumac. These are all low-water-

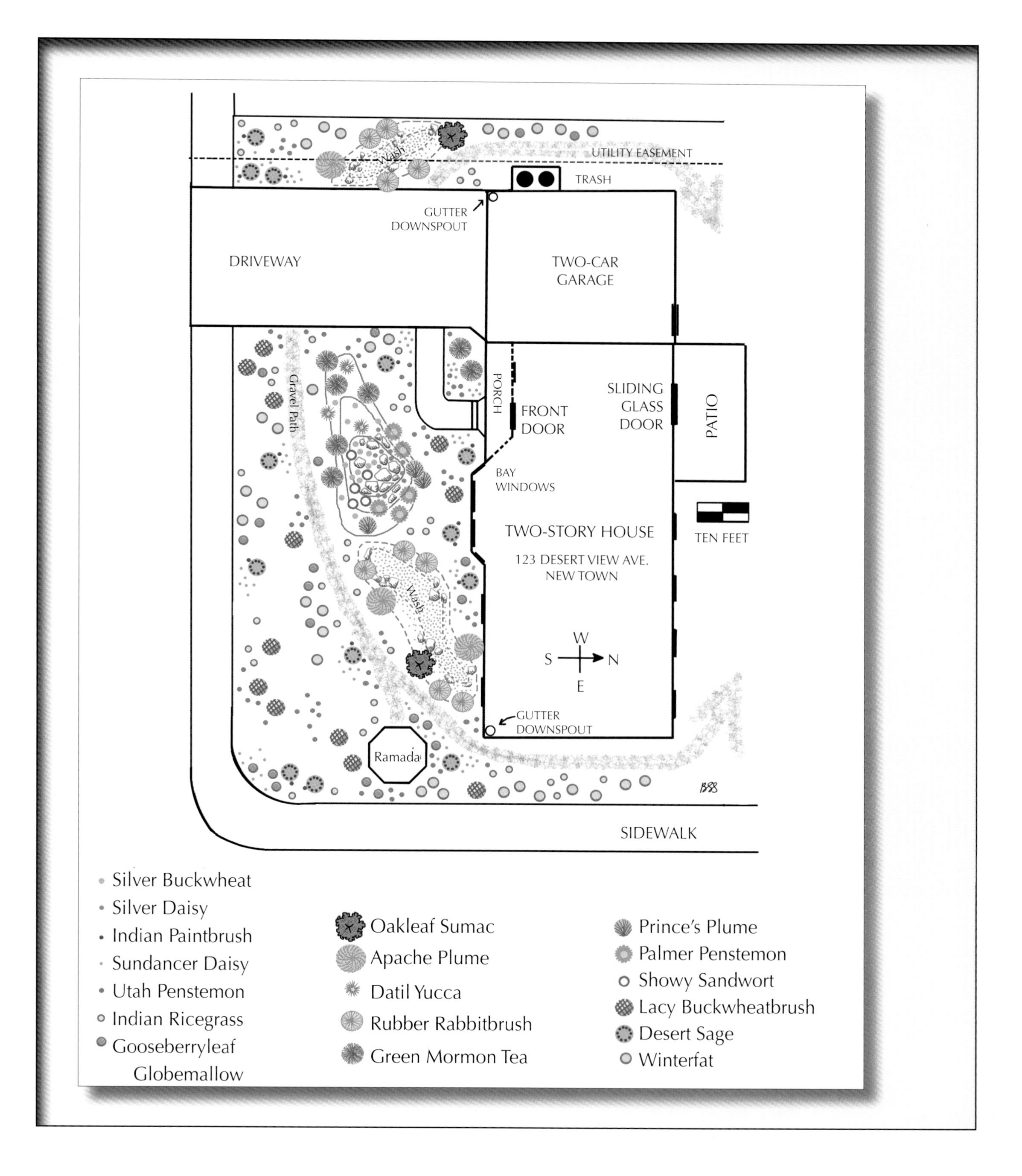
UTILITY EASEMENT
Wash
TRASH
GUTTER DOWNSPOUT
DRIVEWAY
TWO-CAR GARAGE
Gravel Path
PORCH
FRONT DOOR
SLIDING GLASS DOOR
PATIO
BAY WINDOWS
TEN FEET
TWO-STORY HOUSE
123 DESERT VIEW AVE.
NEW TOWN
Wash
W
S
N
E
GUTTER DOWNSPOUT
Ramada
SIDEWALK
Silver Buckwheat
Silver Daisy
Indian Paintbrush
Sundancer Daisy
Utah Penstemon
Indian Ricegrass
Gooseberryleaf Globemallow
Oakleaf Sumac
Apache Plume
Datil Yucca
Rubber Rabbitbrush
Green Mormon Tea
Prince's Plume
Palmer Penstemon
Showy Sandwort
Lacy Buckwheatbrush
Desert Sage
Winterfat

use plants that should do well in a desert landscape as long as they receive some harvested water. They are also fast-growing shrubs that will fulfill their role in the mature landscape relatively quickly.

*Semi-desert Zone Plant Selection Example*

Our third plant selection example will be for a landscape in the semi-desert water zone, where so many suburban dwellers in our region make their homes. Suppose that you live in a subdivision with some pretty strict rules governing the landscaping that surrounds your home, particularly the front yard "commons" that is visually continuous with adjacent yards. The chief expectation is usually that the landscape will look neat and well-maintained. Most of your neighbors have met this expectation using traditional lawn-centered plans, but you want to design an attractive, water-conserving landscape that reduces your "footprint" on the planet and also features many of the plants that characterized the local shrub steppe plant community that was present before development. You also would like to reduce maintenance time relative to traditional landscaping, so you can spend more time on that shady swing on the covered front porch, admiring your handiwork. How do you go about choosing plants to achieve these goals?

In the semi-desert zone, shrubs form the backbone of low-maintenance landscapes. Especially in the front yard, a relatively formal planting of shrubs integrated into a design with fairly extensive hardscape will go a long way toward creating a coherent landscape that meets the tidiness expectations of your neighbors. These mass plantings of shrubs—for example, big sagebrush, fernbush, or green Mormon tea—can be broken up with areas that include smaller plants with contrasting textures and structure. These beds could incorporate a signature succulent such as dwarf or datil yucca; a statuesque bunchgrass, perhaps alkali sacaton grass or basin wildrye; and perennials such as Hopi blanketflower, Utah sweetvetch, Bridges or littlecup penstemon, and dwarf goldenbush. The perennial flowers provide varied forms, foliage patterns and seasons of color, while the succulents and grasses provide year-round structure and textural interest.

To keep a semi-desert landscape within the bounds of neatness, it is usually best to keep perennial plantings fairly small and well-defined, rather than aiming for the "meadow" look. Meadows are among the most difficult native landscapes to maintain in a tidy condition, and are more appropriate in a less formal setting. Resist the urge to try to substitute a meadow for the more traditional lawn area. You will be better off with extensive formal paths that make clean boundaries between the planting areas. You might want to place a planting that features a native grass in an area adjacent to the neighbor's lawn, so that the transition from flat, bright green turf to robust, mostly gray-green shrubs is somewhat softened.

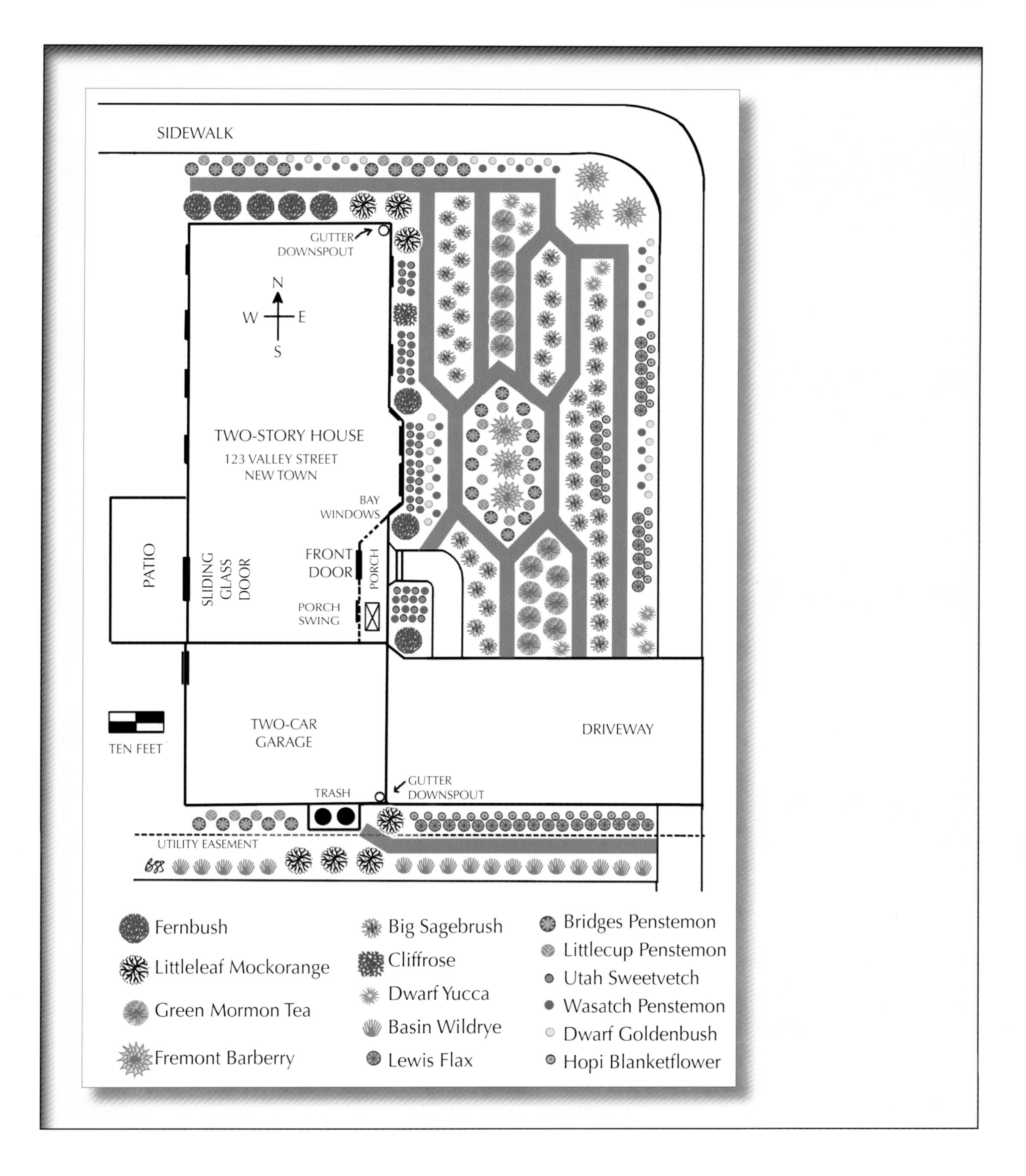
SIDEWALK
GUTTER DOWNSPOUT
N
W
E
S
TWO-STORY HOUSE
123 VALLEY STREET
NEW TOWN
BAY WINDOWS
PATIO
SLIDING GLASS DOOR
FRONT DOOR
PORCH
PORCH SWING
TEN FEET
TWO-CAR GARAGE
DRIVEWAY
TRASH
GUTTER DOWNSPOUT
UTILITY EASEMENT
Fernbush
Littleleaf Mockorange
Green Mormon Tea
Fremont Barberry
Big Sagebrush
Cliffrose
Dwarf Yucca
Basin Wildrye
Lewis Flax
Bridges Penstemon
Littlecup Penstemon
Utah Sweetvetch
Wasatch Penstemon
Dwarf Goldenbush
Hopi Blanketflower

If your neighbor has a sprinkler system that consistently oversprays onto your property, you can take advantage of this extra water to plant species with somewhat higher water requirements. You can also use microsites near eaves-troughing downspouts to provide a little extra water for these species. Fairly fast-growing screening shrubs might be useful in this context, and they can add some mid-height mass to your landscape. A good choice in this situation would be littleleaf mockorange. It can survive even in the low water zone, but will grow much more vigorously with extra water.

If you have issues with small people cutting across your property or want to attract a variety of native pollinators, Fremont barberry may be the shrub of choice for a mass planting on a street corner or along a property line. If your design calls for a vertical accent, you can use cliffrose, which thrives in the semi-desert zone without any supplemental water and has a tall, rugged growth form. Site this plant where you can enjoy its sweet, spicy scent.

You can use a similar strategy to choose lists of species for any plant community on your landscape plan. Think about structure, form, color, and texture and how they combine to provide a balance between complexity and coherence. Read carefully through the description of a plant you are thinking of using, to make sure it has the right plant attributes and cultural needs to fill the niche you have in mind. By the time you are finished, you will have plant lists for each of the native planting areas in your landscape design.

## Preparing the Planting Plan

At this stage of the design process, you know the size of each of the planting areas, and the water zone and plant community that each will represent. You have made a list of the species that you will include in each planting area and have designated them as members of the canopy layer, the mid-height layer, or the understory layer, as appropriate. Now it is time for the most interesting part of the design process, actually deciding how the planting will be laid out on the ground. This involves mapping out how the plants and hardscape features will be spatially arrayed for each planting area. To carry out this design process effectively, you will need once again to tap into your "inner designer." You will need to develop the ability to visualize how a planting will eventually look by working with a layout, or plan, view that shows how the plants will be placed in the landscape. Your final product will be a map of each planting area that shows the placement of individual plants of each species, along with a list that includes a tally of the individual plants of each species that you will need.

The first step in preparing the planting plan is to make another overlay for your base map, this time showing the planting areas and any relevant features included within them. You will need to put a lot of detail onto this overlay, so increase the scale as much as possible. You might want to divide the base map into sections showing individual planting areas, and to enlarge these sections individually. You need to keep track of what you have done to the scale, though. Make sure you have a scale indicator that shows, for example, five feet, on each of the expanded sections. This is important, so that you can make the diameters of the plants you symbolically place onto the planting plan match the scale of the plan. Enlarging all the sections to the same degree will make life a lot easier. Alternatively, you can just enlarge the whole base map and print it on super-sized paper.

*Planning Plant Placement*

There are various ways to proceed with the development of the planting plan at this point, but we like the "paper-doll" approach. Using this method, you first make several to many copies of scaled, labeled circular cutouts for each species. Construction paper or index cards are stiff enough to work well for this. For example, if your enlarged plan has a scale of four feet to the inch, then plants with a mature diameter of four feet would be represented by circles an inch across, plants two feet in diameter would be represented by circles a half inch across, and so on. As you work, it is easy to make more of these circles as needed. Plant diameters at maturity are given in the plant attribute table for each species in the Plant Palette. You can use code letters on the cutouts for each species and even use color-coding as well, to make species' differences easier to see. The advantage to using cutouts is that you can try many different configurations before deciding on the one you like, without having to erase as you make changes. If you know your way around computer graphics software, a "virtual paper-doll" design process also works well and has the same advantages as the hardcopy version described here.

Suppose you chose a mountain brush community dominated by bigtooth maple for a medium water zone in your landscape. Your species list includes bigtooth maple and Gambel oak as canopy species, mallowleaf ninebark, Utah serviceberry, and littleleaf mockorange as mid-height mass, and an understory of shining muttongrass with interspersed drifts of Utah sweetvetch, sundrops, and Lewis flax. Based on the plant attribute table, the maples would be represented by circles scaled to twenty feet and the oaks by circles scaled to fifteen feet in diameter. The serviceberry circles would be scaled to eight feet in diameter, and the mallowleaf ninebark and littleleaf mockorange circles would be scaled to six feet. Each of the understory species would be represented by circles scaled to one foot. Making the muttongrass circles green, the flax circles blue, the sweetvetch circles pink, and the sundrops circles yellow would make it easy to distinguish them in the design. Your planting area is a roughly elliptical area fifty feet long and thirty feet wide, for a total area of about 1,200 square feet.

Now is the time to remember the native landscape that inspired your design, especially the aesthetic features that give it the special feeling that you want to create. Start with the canopy layer. It is OK if the mature crowns of the trees overlap, especially for trees of the same species. A common look in the mountain brush community is a pattern with multi-trunked clumps of trees and fairly wide interspaces. Two trees planted with their crowns overlapping will tend to grow into the shape of one tree with two trunks, quite appropriate for this community. To give the feeling of rivers of space flowing through the clumps of trees, create masses of canopy cover over about 50 percent of the area. Then place the shrubs as mid-height mass along the edges of the canopy. Don't try to crowd in too many plants. Two or three maples, one or two oaks, and one to three of each of the shrub species would probably be plenty for this space. Try placing cutouts of the trees onto the plan, and you will see what a substantial fraction of the area they will eventually cover. But it will be many years before these trees reach mature size. Keeping an area around the young trees free of understory plants will help them grow faster, but this bare area only needs to extend out a few feet, not over the whole area that the canopy will eventually cover. The remaining canopy area can be planted to understory species.

The shrubs need to be placed fairly far out from the center of the tree canopies, especially the serviceberry, which can be ten feet tall when mature. Move the cutouts of the trees and shrubs around until you have a pattern that pleases you. Think in terms of the design principles of coherence versus complexity, mystery versus legibility. By using a limited number of species and by placing multiple individuals of these species in a repeating pattern in space, you give the design coherence. On the other hand, by using several species rather than a single species, you increase the complexity and interest. Complexity is also increased by using multiple layers to create structure. Both legibility and mystery are served in the design by creating open spaces around and through the shrub and tree plantings. A solid thicket of trees and shrubs looks impenetrable and uninviting rather than mysterious and beckoning. This is what you will have if you fail to take the mature size of the plants into account and try to cram too many trees and shrubs into the planting area.

Once you have a placement that you like for the trees and shrubs, work on the understory layer. You will obviously need a lot more copies of plant circles for these species, but, again, open space is good. In general, leave spaces between understory plants that are roughly the size of the plants themselves, for a total cover of around 50 percent in the planted areas. Do not place the plants evenly, either in rows or on a grid. Deliberately vary the spacing, with some plants closer together and others further apart, for a more naturalistic feel. You could even leave slightly wider areas that are not planted as informal winding paths to further increase both symbolic legibility and actual physical access to the planting. Use the shining muttongrass as the matrix species, and introduce the flowering perennials into the matrix in patches or drifts that taper off from dense centers and interfinger with one another. This looks more coherent, and also more natural, than just dotting the flowers around at random through the grass planting.

If the planting area includes any hardscape features, such as landscape rocks or an informal seating area, be sure to consider these in the planning process. Don't forget to take any topographic relief into account, as it will affect both the microenvironments for individual plants and the visual aspect of the planting as a whole. For example, if the planting is on a steep slope, make sure that any physical access path traverses the slope, the way a deer trail would, rather than charging straight up from bottom to top. This would be an invitation to erosion as well as being visually unattractive and uncomfortable for walking.

*Visualizing the Plan*

When you have a design that you think you like, try visualizing what the planting will look like when mature. Pick a hypothetical vantage point, such as the front walk or the living room window, and imagine the way the tree and shrub masses will look in relation to the openings. If you can sketch this on paper, so much the better. Remember, this is not art. This is just a temporary design device for your eyes only. Does the design look attractive? Does it have a pleasing balance of complexity and coherence, of legibility and mystery? Does it look like a place that would welcome you to enter? Does it capture the visual essence of the plant community you are trying to create? Using this process can help you detect problems, such as noticing that there is really no visual access to the landscape once the planting is mature. Or perhaps the planting is too sparse, with more of a savanna feel than the cozy mountain brush atmosphere you were trying to create. Go back to the paper doll version, and make any changes suggested by this visualization process. If you like, you can repeat this visualization process as many times as necessary, until the design feels right.

When you are satisfied with the design, trace the circle for each individual plant onto the base map and label it by species. You can save out the paper doll circles that you remove as you do this, then use them to tally the total number of individuals of each species that you will need for the planting.

The process for preparing the planting plan for any area will be similar to the process described above, regardless of what plant community you are trying to create. In general, plantings representing drier water zones should include more unplanted area than those representing wetter water zones. Fifty percent cover is about right for the foothill water zone, whereas seventy to eighty percent cover might be appropriate for a mountain planting, and cover as low as twenty percent might be about right for a minimal-water planting. Again, the cover should not be uniformly distributed, but instead should reflect natural patterns based on underlying or designed microhabitat variation. In desert communities, for example, the plants often occur in multi-species clumps with wide interspaces. This could be because an exceptionally good microsite can support many inhabitants. Or it could be that the presence of a mature plant of one species can facilitate the establishment of another, perhaps by acting as a nurse plant that shades the young seedling. The net result is a pattern that is far from either random or regular, and a minimal-water landscape will look and feel more realistic if some of the plants show this kind of spacing.

If you have multiple planting areas that represent the same plant community, make sure that the planting plans for these areas are prepared with careful reference to each other. Multiple areas that repeat the same suite of species with minor variants will visually integrate these areas across the landscape and greatly increasing the feeling of place that is generated.

*Mountain lover and Bigtooth maple*

# How to Water Native Landscapes

Chapter Three

The next step in the process of designing and installing a native landscape is to deal explicitly with how your plantings will be watered. Your completed planting plan shows the species that are to be planted into each area, as well as the spatial configuration of the planting. Each planting area has been designated as belonging to one of the five watering zones: minimal, low, medium, high, or very high. Now you will decide the most efficient and best way to provide water to each watering zone. Plants in all zones will need to be watered during establishment, so there must be some way to get supplemental water into each area, at least initially. But only areas representing watering zones that will receive supplemental water on a regular basis will need more permanent irrigation systems. These are the zones that represent plant communities from places that receive more annual precipitation than your site and that cannot be situated in sufficiently favorable microsites to preclude the need for added water.

There are four principal ways to water plants: water harvesting, hand watering, sprinkler irrigation, and drip irrigation. Which method you choose for a particular water zone will depend on how often the area will likely need to be watered, how much water will need to be added, and how the plants are spaced. The first method, water harvesting, uses water received onsite, while the other methods involve the use of water from offsite, usually water from the same supply line that provides culinary water to your house. Another possible source of offsite water that is sometimes available is ditch irrigation water—a remnant of farming days in small towns of our region. Or in some scenarios, "gray" waste water (wash water) from the house may be an option—see the resource section for more information.

# Water Harvesting

Water harvesting is a much-neglected method of irrigating landscape plantings. It uses water from the rain and snow that fall onto non-planted areas of your site to supplement the water available in planted areas. It involves collecting rainwater or snowmelt from an impervious catchment area and redistributing it to an area where it can be used by plants. Sometimes a storage system, such as rain barrels or a cistern (underground water storage tank), is included to make it possible to water plants during periods when the catchment areas are dry. Water-harvesting systems can be simple or complex. Large landscape rocks can act as catchment areas that redistribute water to the areas around their bases. This, combined with the shade created by the rock and its ability to lower evaporation from the soil surface, makes places adjacent to landscape rocks distinctly wetter than surrounding areas. Another simple form of water harvesting is laying a new slab of cement for a driveway with a slight slope toward the adjacent planting area, so that rainwater will run off into the planting. Even very shallow slopes are sufficient to create flow over a smooth surface.

The catchment area represented by the roof of the house is usually the most significant source of extra water for the landscape. And even though water harvesting is not 100 percent efficient, the amount of extra water that can be made available to plants is considerable. Most houses have gutters and downspouts to collect water from the roof and carry it away from the house foundation. It is not difficult to intercept the water at the base of the downspout and direct it to the designated planting area, as long as the pipe or drain that carries the water slopes downward. Again, only a shallow slope is required, about one foot of drop for every hundred feet traversed. An elegant way to transport the water is to use a French drain, which is essentially a gravel-filled ditch, often lined with polyethylene plastic sheeting. These structures are commonly used to divert water away from areas where it is not wanted. But they are just as good at carrying the water to areas where it is desired, namely the planting areas in your design that will depend on harvested water for part of their water needs. The planting area receiving the water can be shaped so that it holds the water, either in a depression or swale, or on a flat surface with berms to make a "moat." Often the areas receiving the water are arranged in a series, each slightly lower than the next, so that once the water has filled the first receiving area, it can spill gently over into the next one. It is also possible to store water harvested from a roof and distribute it later using drip irrigation. We include some excellent references in our resource section to help those of you interested in water harvesting to engineer a harvesting and distribution system that will make you proud.

## Water Harvesting

You can take advantage of water harvesting on a small scale by planting species with higher water requirements around landscape rocks, especially on the shady north side. The rock sheds water down its sides and also protects the water underneath from evaporation.

A French drain is a clever device for carrying water some distance away from a gutter downspout. A slightly sloping trench is filled with gravel, and the water flows through this gravel to its destination. The trench is often lined and covered with plastic, so that it conducts water more efficiently and can be buried in soil without clogging the gravel.

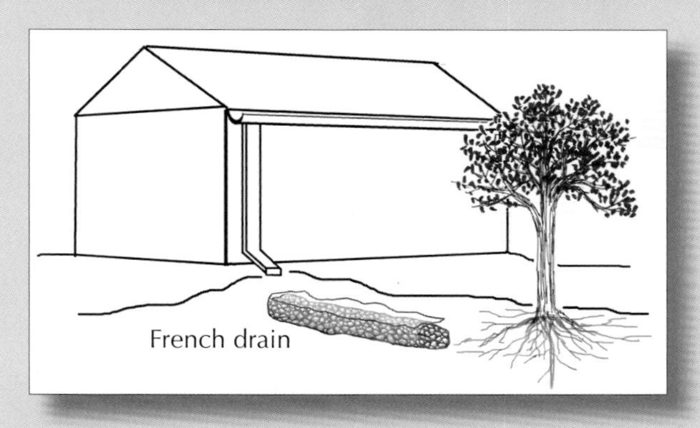


Roof water can be guided directly from the gutter downspout to adjacent shallow planting basins. These can be tiered, so that the lower basin receives harvested water only when the upper basin fills to overflowing.

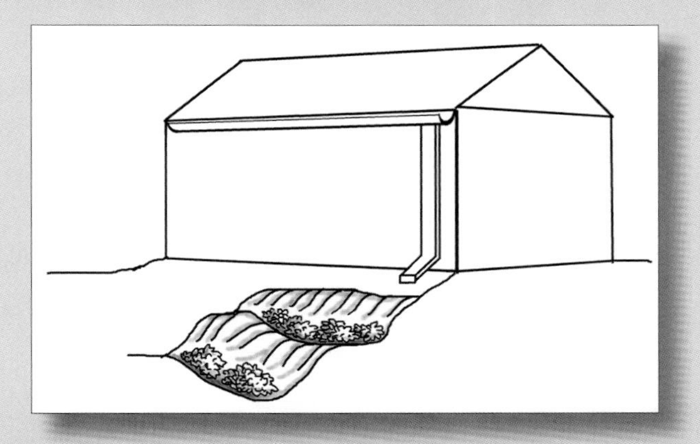

It is important to note that in certain areas of the Intermountain West, state or local ordinances restrict some forms of water harvesting and some uses of gray water. Be sure you know what is permitted in your area before you begin planning your own system.

### Watering by the Numbers

### *1. Estimating Water Yield from a House Roof*

Every inch of rain that falls onto your roof drops 0.625 gallons onto each square foot of roof surface.

To calculate the potential water yield per inch of rainfall:
***house length × house width × 0.625 = gallons per inch of rain.***
For example, if your house is 80 feet long and 40 feet wide, a 1″ rainstorm can generate 80 × 40 × 0.625 = 2,000 gallons of water.

To estimate the total potential water yield from your roof:
***gallons per inch × average annual precipitation = total gallons.***
In a 12″ precipitation zone, your potential water yield could total 2,000 × 12 = 24,000 gallons of water over the course of the year, potentially a rather substantial contribution to landscape irrigation.

## Adding Supplemental Water

Hand watering, sprinkler irrigation, and drip irrigation are all methods of adding supplemental water from offsite. Hand watering is usually most appropriate during the establishment phase. It can also be used during exceptionally dry years to apply supplemental water in zones that usually do not require extra water. Watering the wells around trees by filling them from a hand-held hose is hand watering, as is carrying around a watering can or milk jugs to water new transplants in an established planting that does not generally require supplemental water. If you water small plants with a hose, use a spray wand to keep from washing out the roots. Occasional hand watering is not a big job unless the area is large, though it can be somewhat labor intensive.

Sprinkler irrigation is the traditional method for watering lawns, and is best suited for watering large areas that are densely planted. It can also be used as a temporary method of providing supplemental water during exceptionally dry years in zones that do not have permanent irrigation systems. Using a portable sprinkler on the end of a hose to water a large area is essentially an extension of hand watering. In contrast, turf sprinkler irrigation systems, with buried supply lines, multiple sprinkler heads on a grid, and automated run times, are at the other extreme, requiring little labor or even thought once they are installed and placed in operation.

Sprinkler irrigation has several disadvantages in the context of native plant landscaping. First, it wastes water, because a significant fraction of the water is often lost to evaporation before it even hits the ground. Second, it applies water equally to plant root zones and interspaces, so that, unless the plants are very closely spaced, it tends to encourage weeds. Third, many native species are prone to leaf diseases when their foliage is wetted on a regular basis. And fourth, it is difficult to water uniformly if there are plants present that are tall enough to obstruct the spray. Design and installation of overhead sprinkler systems is therefore not covered in this book. Fortunately, if you already have a sprinkler irrigation system in place, it is not too difficult to switch this system over to a drip irrigation system, the method of choice for watering native plant landscapes.

Drip irrigation is characterized by the use of low pressure to apply a small volume of water over an extended period of time directly to a localized area around the plant. Applying water slowly over a targeted area works very well in native landscapes, because it waters only the root zone, not the interspaces. Very little water is lost to evaporation, and the problems of spray interception and diseases associated with wet leaves are avoided.

## A Primer on Drip Irrigation

Drip irrigation is an easy concept to grasp, and installation of a drip irrigation system can be a relatively simple process. The exact design of the system is not critical, so do not worry too much about calculating everything perfectly. But you will probably approach your installation project with more confidence if you have an underlying understanding of some drip irrigation principles. This book can only provide the basics of drip irrigation, but you can find sources of additional information listed in the resource section.

### *Drip Layout*

In drip irrigation, water is provided from a supply pipe that is usually at high pressure (50–80 psi, or pounds per square inch), such as a hose bibb or converted sprinkler system. You first need to reduce this pressure to 10–30 psi, the low pressure needed to operate a drip system, by installing a pressure regulator in the line. Next, this water at reduced pressure is passed through a filter. This is necessary because the orifices of the drip emitters, the devices that actually apply the water, are very small and easily clogged. Even culinary water can contain particles that can clog the emitters. The filtered water passes through a fitting that connects the filter to the drip line itself. This drip line, or tubing, is usually made of flexible polyethylene (PE) and is usually half-inch tubing (actually 16 mm or 0.62" in diameter). The emitters are installed at intervals along this tubing, which is capped at the end to prevent a loss of pressure to the emitters.

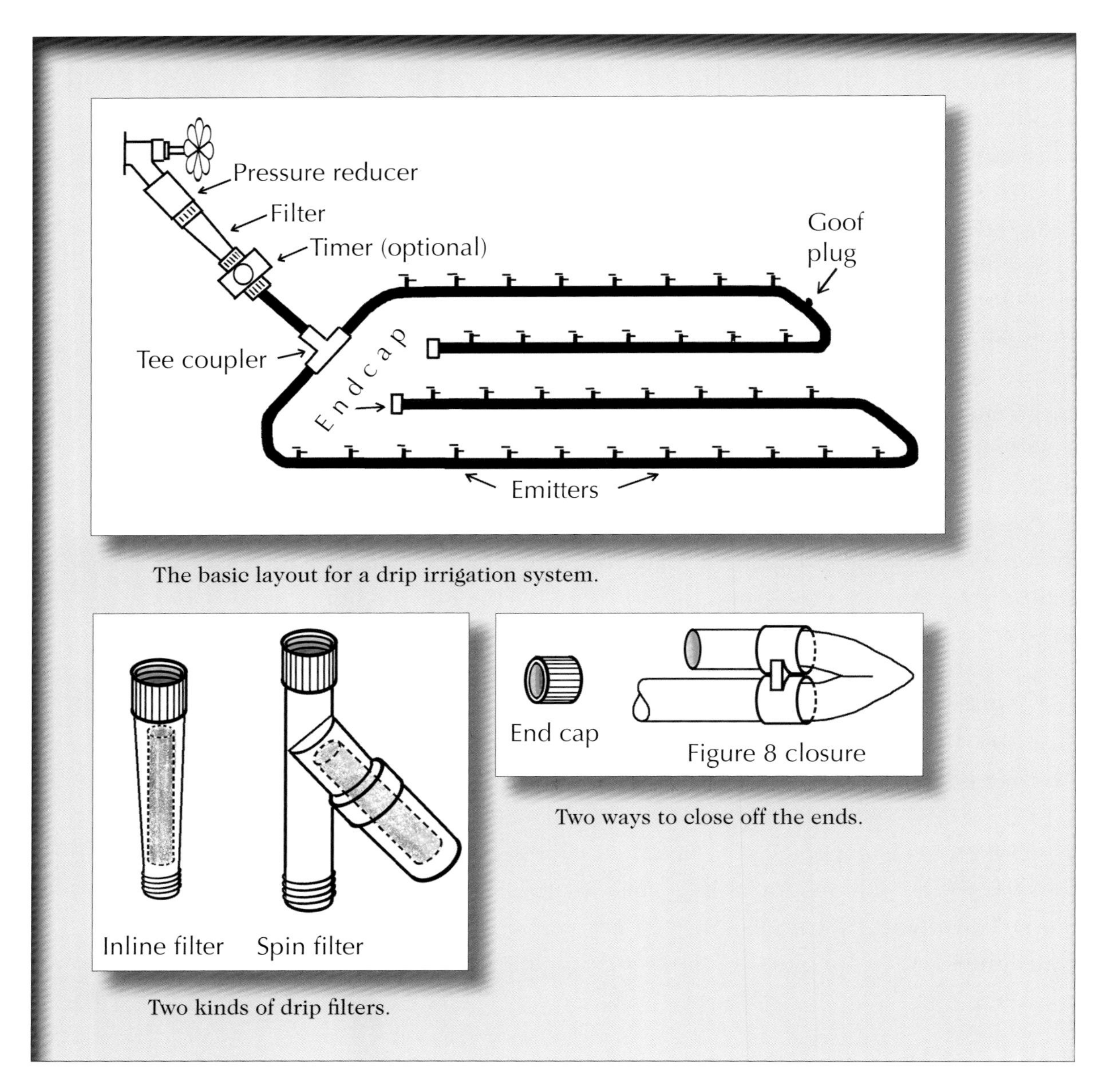


The basic layout for a drip irrigation system.

Two ways to close off the ends.

Two kinds of drip filters.

This is the simplest case scenario for a drip system, with a single drip line attached directly to a hose bibb. In fact, many variants are possible. One alternative is to install the drip system from scratch in a manner analogous to the installation of a sprinkler irrigation system, with buried lines, usually made out of rigid polyvinyl chloride (PVC) pipe, and multiple risers where drip lines are then attached. Installation of this kind of system involves a lot of serious trenching and leveling, and is probably best left to the same professionals who routinely install sprinkler irrigation systems, though a zealous do-it-yourselfer could do the job. If you hire professionals to do the installation, you need to make sure that they understand exactly how the drip system is designed and how it is intended to function.

Another common drip irrigation variant, one that is superior in design to a single long drip line attached to a water supply, is a design that has a main line that branches one or more times. This makes it possible to cover more area with the drip system, and to provide different watering regimes to different parts of the system. A branched system is also more efficient, as will be explained below.

A third drip irrigation possibility is to retrofit an existing sprinkler system as a drip irrigation system. Because sprinkler systems operate at high pressure and the sprinkler water passes through large orifices, these systems do not need filters or pressure regulation. To convert a sprinkler system to drip, it is necessary to insert both a pressure regulator and a filter into the line. Some drip conversion fittings have built-in pressure regulation and filters, but for other, simpler types of fittings, these components must be inserted into the supply line. The sprinkler heads can then be replaced with drip lines. Any sprinkler head locations that are not needed can just be capped off.

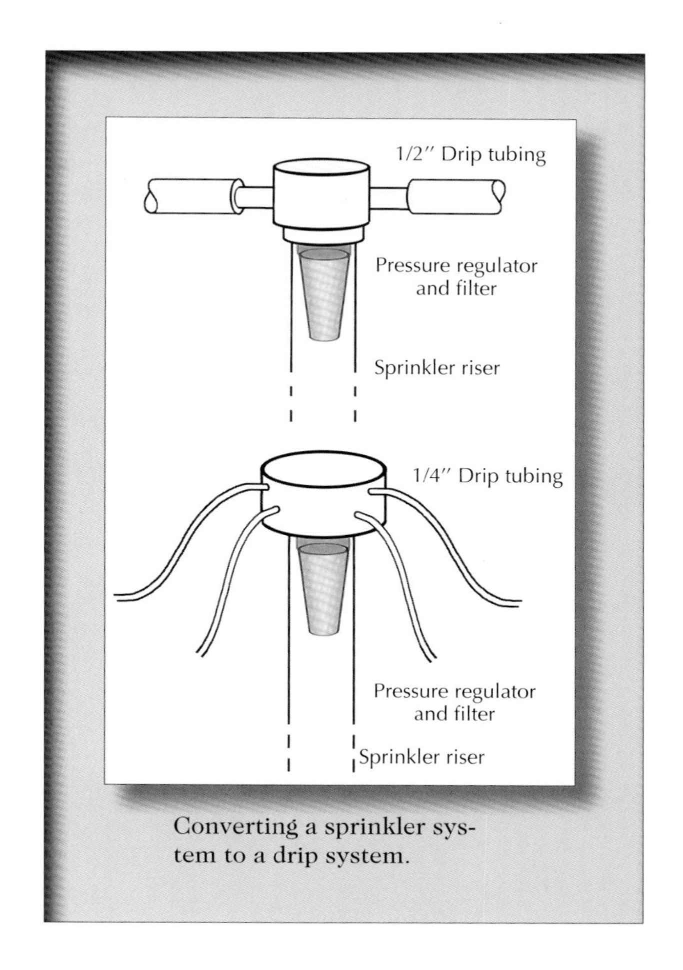

Converting a sprinkler system to a drip system.

The number of emitters that can be installed in a drip system depends on several factors. The water demand created by the emitters depends on the number and flow rate of the emitters, while the ability of the system to meet that demand is determined by the size of the supply pipe. If the demand exceeds the ability of the supply pipe to provide water, the emitters toward the ends of the lines will either not drip at all or will drip at a lower rate. Because all the water in the supply pipe (e.g., the pipe that supplies your hose bibb) is reduced to the same low pressure, the supply of water to the emitters will depend directly on the diameter of the supply pipe. It would theoretically be possible to force a small supply pipe to provide the same amount of water as a large supply pipe, but only by increasing the pressure and therefore the flow rate beyond the tolerances of the drip irrigation components. Selection of appropriate emitter sizes is covered below. If you need to have more total emitter output per hour than your supply pipe can provide, you can split your system into two or more subsystems. By running these subsystems at different times, you will be able to supply water to as many emitters as needed. There are other advantages to multiple subsystems as well, such as providing the correct timing and amount of water to plantings in different water zones.

## Watering by the Numbers

### *2. Calculating Water Supply/Demand for a Drip System*

The flow rate of the emitters in gallons per hour (gph), multiplied by the number of emitters, determines the water demand in gallons per hour on a drip line.

In general, a 3/4″ supply pipe can provide water to 200–300 1–gph emitters, while a 1″ supply pipe can provide water to 400–700 1–gph emitters. The lower numbers apply to a drip system attached directly to a hose bibb and therefore to house plumbing, while the higher numbers apply to an independently installed system.

The important number is the total number of gallons. For example, if you use 2–gph emitters, the ¾ inch supply pipe can only support half as many of them as when you use 1–gph emitters, namely 100–150 emitters, whereas if you use 1/2–gph emitters, the 3/4″ supply pipe can support twice as many, or 400–600 emitters.

Another factor that affects how well the emitters at the end of the drip line will function is the length of the drip line. This is because the interior wall of the line creates friction as the water passes through, and this decreases the pressure, so that by the end of a long run of line, the pressure could be too low for the emitters to operate correctly. Two hundred feet is considered the maximum acceptable length for a single run of drip line, but if you can reduce the length of the run, the drip system will probably operate more precisely, especially on irregular terrain. The way to do this is to branch the lines. With a branching system, the maximum run length can be kept shorter while still maximizing the total length of line. The limiting factor in a well-designed system is the number of emitters at a given flow rate that the water supply can support, rather than the length of the run.Sloping ground can also cause uneven pressure along the length of a drip line run. If the ground is sloping, run the main line up or down the slope and the lateral lines along the contours. If possible, loop your main lines so that they return to a spot near the starting point. This helps equalize the pressure to emitters throughout the line and is generally a good idea, even on apparently level ground.

## Watering by the Numbers

### *3. Reducing Friction Loss in the Supply Lines*

Friction loss in a branched line is determined by the length of the run, which is the distance from the beginning of a line to the end of its longest branch. If the line is attached to the water supply in the middle of a 400–foot run, for example, then in effect you have two lines, each 200 feet long, the maximum recommended run length.

Or suppose you run your line out 130 feet, then create 8 laterals at the end at 10–foot intervals, each 50 feet long. Your longest run is 180 feet, namely the sum of the length of main line to the point where the last lateral is attached (130 feet) plus the length of the last lateral (50 feet). But you have supplied 400 feet of drip line to your plants.

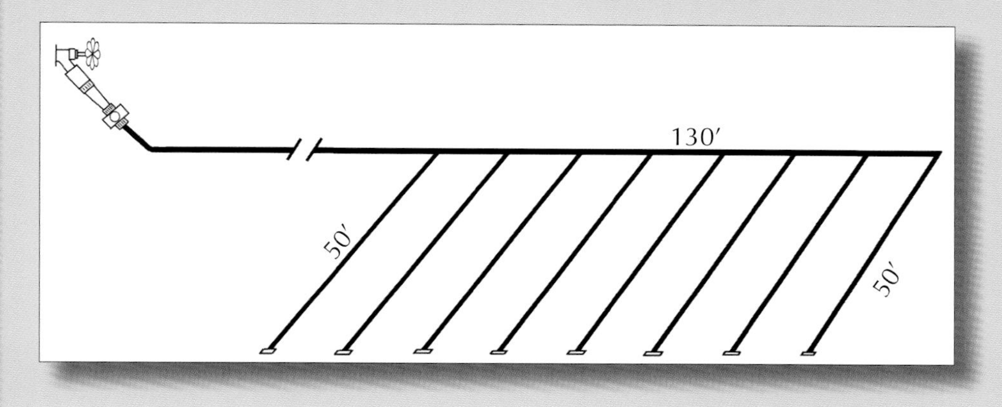


*Emitter Basics*

There are a number of different designs for emitters, and each is suited to a different application. The three common designs are point-source emitters, inline emitters, and microsprinklers.

Point source emitters are the original form of drip irrigation, where a small piece of well-designed plastic restricts the flow of water from a large line down to a drip, so that only a small surface area is wetted. Note that the numbers stamped into the emitter plastic are almost always in liters, not gallons. For example, the numeral 4 means four liters per hour, or approximately one gallon per hour (gph). A point-source drip emitter has a barbed inlet that allows it to grip and hold when it is attached to the line. Pressure compensating emitters, which maintain a constant pressure inside the emitter regardless of pressure variations in the line, perform much better than regular emitters, especially

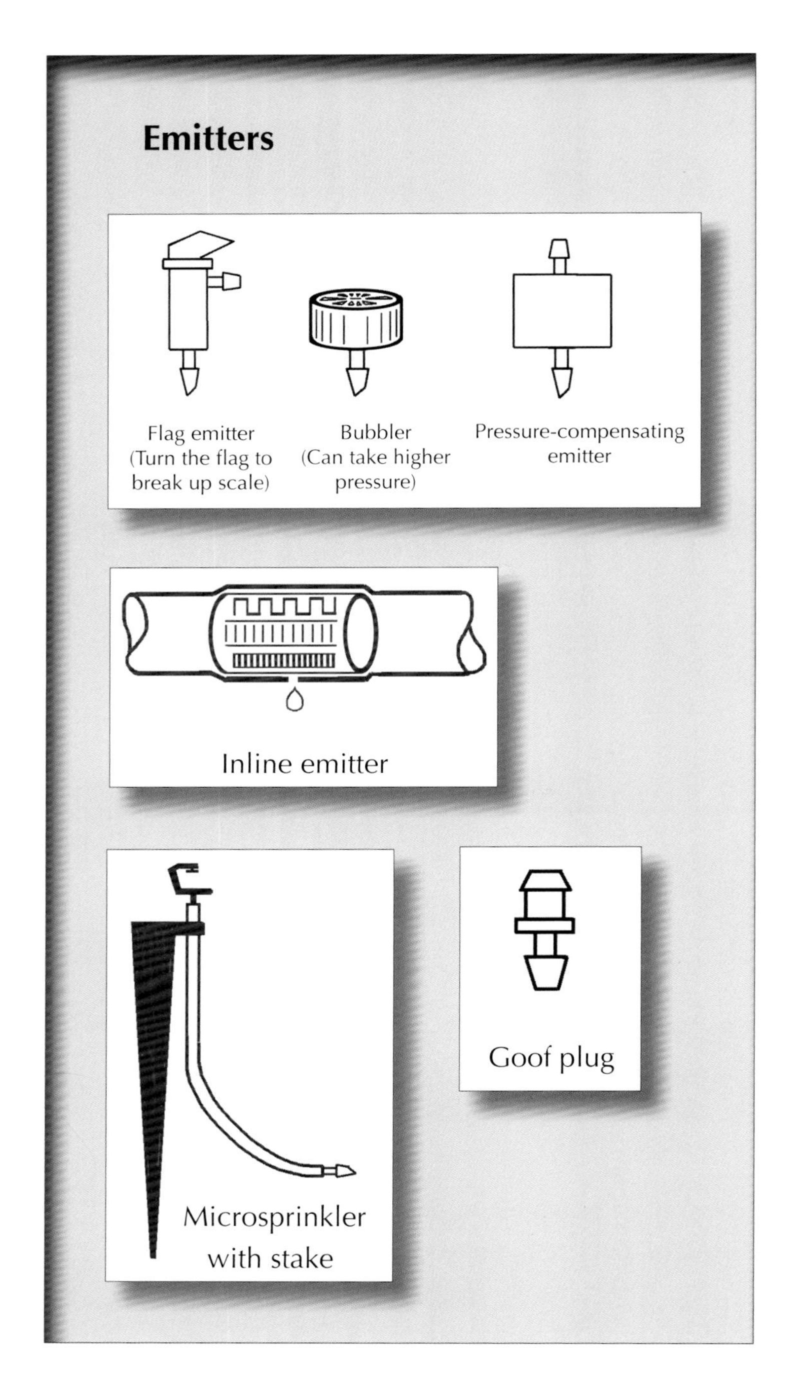


where the ground is sloping. All point-source emitters can drip directly onto the ground. Some emitter types can be made to drip into a narrow piece of "spaghetti tubing" pushed over the top of the emitter. This tubing permits pinpoint application of the water at some distance from the line. Spaghetti tubing is often not necessary, as the lines generally pass close by the plants, and the emitters can drip directly onto the root zone without any extra tubing.

Inline emitters are emitters that are preinstalled inside the line. The best inline emitters are drip emitters inserted into the polyethylene tubing during the manufacturing process. These are spaced evenly along the length of tubing at 6-, 12-, 18-, or 24-inch intervals, and have either 1/2, 1, or 2 gph flow rates. There are two other types of inline emitters, trickle tape and porous tubing, but we do not recommend them. Trickle tape is difficult to handle and not very durable, while porous tubing (soaker hose) is hard to control and clogs quickly, especially where the water is hard. These types of inline irrigation might be economical alternatives for a temporary watering system, but for the most part they should be avoided.

The third type of emitter is the microsprinkler, which is really somewhat of a hybrid between a drip emitter and a sprinkler. Microsprinkler irrigation is relatively low volume irrigation that pushes water through a tube and into a plastic head that breaks up the stream of water into a spray pattern, similar to a spray head in a regular overhead irrigation system. Microsprinklers come in different specifications that deliver water at rates from 2 to 15 gph and cover a circular or semicircular area of ground from one to eight feet in diameter. The spray head is detachable from the tube, and the tube is usually attached to a stake that is pushed into the ground. This tube is made from heavy-walled PE that is more rigid than spaghetti tubing. A barbed fitting similar to the one used with drip emitters connects the tube to the supply line. Microsprinklers are good for larger plants because they irrigate a larger area of the root zone, with more water coming out of a bigger opening than that for drip emitters. Because they are upright and more visible, microsprinklers are easier to troubleshoot, but they are also much more likely to be damaged. They offer little advantage over drip emitters, except for large shrubs and trees.

Another handy little piece of plastic that looks a bit like an emitter is the goof plug. This is used to correct any mistakes when you are placing emitters, that is, to plug up a hole where an emitter has been removed.

*Putting the Pieces Together*

Assembling a drip irrigation system is a lot like playing with tinker toys. The fittings that attach pieces of drip line to each other are called couplers or connectors, and they come in several configurations. Straight couplers attach two pieces of tubing in a line, while tee couplers attach three pieces of tubing, and elbow couplers attach two pieces of tubing at right angles, permitting the line to go around a corner. The couplers that attach directly to PE tubing are generally of two types, compression fittings or barbed fittings. Each has its proponents. To attach tubing to compression fittings, the tubing is pushed into a tight opening at the end of the fitting. Compression fittings that are not properly drained can trap water that can later freeze and cause damage to the fitting. To attach tubing to a barbed fitting, the tubing is pushed over a series of ridges on the outside of the opening into the fitting. Barbed fittings tend to stretch and weaken the end of the PE pipe,

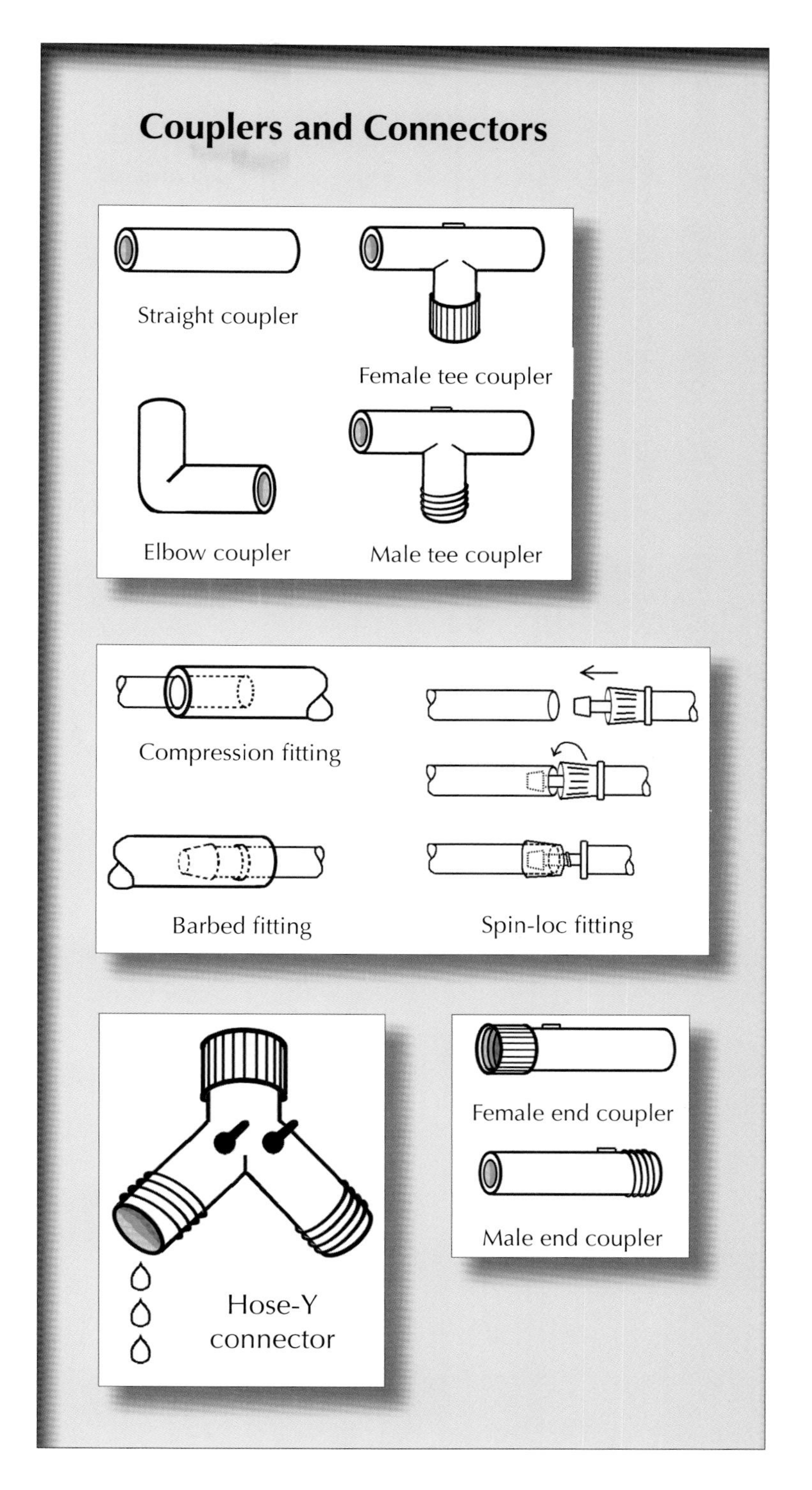

eventually causing it to split. In a third type, a spin-loc fitting, the tube slips over a nipple with an O-ring, which is then tightened down with a twisting nut. It is important to make sure that all connectors are the correct size for the PE tubing. Tubing sizes are actually in metric units, and there are two tubing sizes (16 mm and 18 mm) that are commonly referred to as "half-inch." The fittings for these are not interchangeable.

Some connectors take PE tubing directly at all connection points. Others, called adapter connectors, have at least one male or female threaded connection point for attaching to other connectors, for example, when interfacing with a line of a different size. For threaded connectors, it is important to note whether they have pipe threading or hose threading. Pipe threading is finer than hose threading, and the two are not interchangeable. Many connectors also have built-in filters to further reduce the chances of clogging the emitters.

Another very important fitting for drip irrigation systems is the valve. A valve is a device for regulating the flow of water. In a system attached to a hose bibb, the valve (faucet) that turns on the water can also regulate the amount of water that passes through; in effect, this is a manual form of pressure regulation. In most cases, though, the valve will either turn the flow on or turn it off. If you have a system with multiple lines and a total number of emitters that exceeds the capacity of your supply line, it will be necessary to install valves on each subsystem, so that water can be delivered to the different subsystems at different times. The valves could be manually operated ball valves, similar to those used on hose Y-connectors, or they could be sophisticated electronic valves. Areas that represent different water zones in your landscape will need to be on separate subsystems.

Another use for valves is in connection with automatic timers. The timer signals the valve to open or close at pre-set times, thus controlling the timing and duration of the irrigation. This is the method commonly used with sprinkler irrigation systems. A problem with automatic timers for drip irrigation is that they rarely have options for irrigation intervals and durations that are appropriate for a drip system. Make sure when you buy a timer for your system that it has sufficient programming flexibility to meet your needs—most are designed for watering lawns. The timer program needs to be able to specify watering periods of one to eight hours at intervals at least up to thirty days. Eight hours may seem like a very long time, but remember, the water is being applied *slowly*. Many a plant on drip irrigation has died from overwatering because the timer insisted on fifteen-minute irrigations every other day. This may conceivably add up to the correct amount of water, but it is not the correct way to apply it, for reasons that are discussed below. It may sometimes be easier to dispense with the automatic timer altogether and operate your drip system manually.

### Watering by the Numbers
### *4. Gallons Versus Inches*

The approximate area watered by a single point-source drip emitter equals a circle with a radius of 1 foot. The area of a circle equals the radius squared times *pi* (3.14):

**area watered: *1 × 1 × 3.14 = 3.14 square feet (or approx 3 square feet)***

The amount of water needed to apply 1″ of water to a square foot is 0.625 gallons. To add 1″ of water to the 3 square feet watered by a single emitter:

***0.625 gallons × 3.14 square feet = 1.96 gallons (or approx 2 gallons)***

*To apply 1″ of water, the emitter needs to apply about 2/3 gallon × 3 square feet (or approximately 2 gallons of water).*

This has the added advantage of providing an opportunity to exercise judgment on when the water is applied. How many of us have shaken our heads in exasperation upon seeing the neighbor's automatic lawn sprinklers come on in the middle of a downpour?

*How Much to Water*

One thing you will have noticed in this discussion is that drip application rates are expressed in terms of gallons, while the watering recommendations described earlier in this book are expressed in terms of inches. Our task now is to get these two descriptions of water application to "speak" to each other. We first need to know the area of ground that a single drip emitter can water. This is difficult to learn by inspection, because most of the water that drips onto the ground disappears. At appropriate application rates, only a small spot on the soil surface is wetted. Fortunately, other people have made a study of this subject. A frequently recommended spacing between emitters is twenty-four inches, or two feet. The water spreads laterally underground, out of sight, to a distance of roughly twelve inches, or one foot, in a circle that radiates out from the emitter. You could therefore say that, on average, a single emitter can wet an area equal to the area of a circle with a radius of one foot, or a little more than three square feet. To apply an inch of water to the three square feet that it can wet, the emitter needs to apply about two gallons of water.

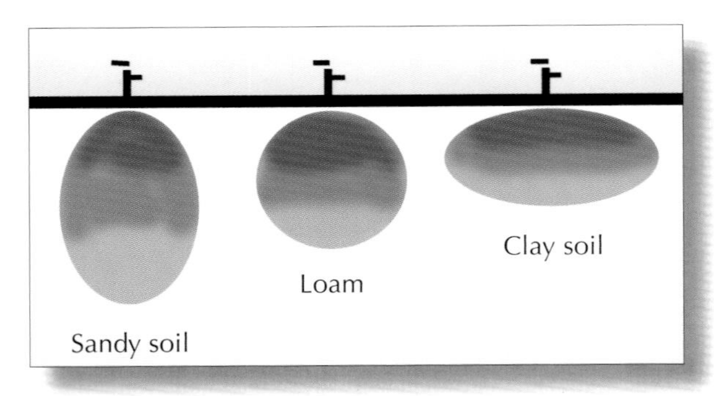


How the water spreads in different types of soil.

The distance out from the drip emitter that the water actually spreads underground depends on soil type. Because of smaller pore size, fine-textured clay soils conduct water laterally through wicking action better than coarse-textured sandy soils, and loamy soils have an intermediate lateral spread. As long as the water is applied at a suitable rate, different soils can absorb water equally well, but in sandy soil the water will penetrate more deeply, while in finer-textured soil it will not penetrate as deeply but will spread more to the sides. Thus the effective area watered by a drip emitter on a sandy soil will be somewhat less than the area watered on a loam or clay soil. The flow rate of the emitter also influences the lateral spread of water in a soil. The lateral spread can be increased by increasing the flow rate, which is appropriate in a sandy soil, where you might want to increase the spread. In heavier soils, slower flow rates are needed to keep the water from puddling on the surface, and this will decrease the lateral spread.

In practice, differences in lateral spread due to soil type or flow rate are not too important, provided that the emitters are arrayed so that they target individual plants. As long as you do not grossly overwater, native plants are likely to send their roots to whatever depth the water has penetrated. They will use the water whether it spreads laterally or penetrates deeply, pretty much minimizing the effect of soil type on water availability. This might not be true for tomatoes or petunias, which have shallow root systems, no matter how deep the water penetrates. For these plants, a sandy soil must be watered more often, because there is less water held in this shallow root zone. But part of the remarkable ability of many native plants to survive with little water involves their capacity to extend roots to whatever depth the water penetrates and to whatever extent it spreads.

This capacity for root development in response to water availability also means that it is not necessary to apply irrigation water over the entire crown area of a plant. This is true even for trees from wetter places. Tree canopies intercept rainfall, so that the ground directly under the tree receives less rainfall than open ground next to the tree. Much of this intercepted water is redistributed in the canopy and ends up dripping from the branch tips in a ring around the outer perimeter of the tree. Surface roots are often concentrated beneath this drip line, so that irrigation water applied in this area is more effective than water applied near the trunk, where surface roots tend to be sparse. This is the reason that drip emitters for trees are usually arrayed in a ring just inside the drip line. As long as a sufficient volume of water is applied within the crown area of a plant, the exact placement of the water is usually not too critical.

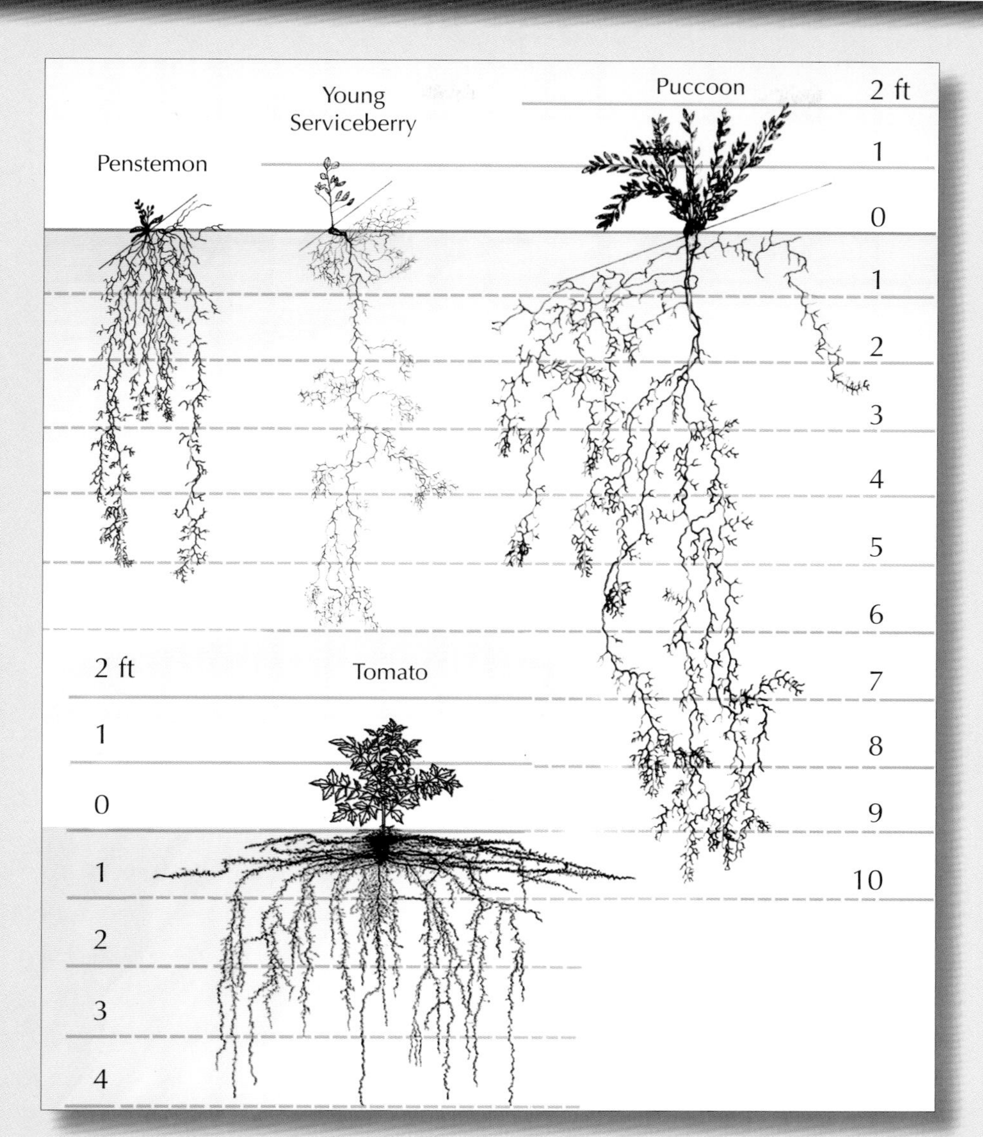


Root systems excavated and mapped by:
Weaver and Bruner 1927 (tomato)
Spence 1937 (natives)

## Natives Versus Tomatoes: Root System Differences

These root systems were excavated and mapped on a hillside in central Idaho over eighty years ago. Native plants have deep and extensive root systems but usually lack a major concentration of roots near the surface. Even these small plants, including a serviceberry shrub hardly past the seedling stage, have roots 5–10 feet deep. In contrast, roots of cultivated plants like tomatoes have dense root systems concentrated near the surface.

### Watering by the Numbers

### *5. Figuring Water Needs from Plant Size*

Gallons needed to apply 2″ of water to a plant depends on its size (crown area). To get the total gallons needed, multiply the crown area in square feet by 0.625 gallons × 2″. For a plant with a crown radius of 2 feet:

**crown area:** ***2 × 2 × 3.14 (pi) = 12.56 (or approximately 12 square feet).***

**gallons needed to apply 2″:** ***12.56 square feet × 0.625 gallons × 2 = 15.7 gallons (or approximately 16 gallons).***

Shortcut: To get the total gallons needed to apply 2″ of water to a plant of any size, just multiply its crown area by 1.25.

*An even easier rule of thumb for the approximate number of gallons needed in order to apply 2″ of water to a plant of a given size is just to take the square of its crown diameter (multiply crown diameter by crown diameter).*

To design your drip irrigation system, you first need a clear idea of how much water you will need to apply to each plant in a single irrigation. *Our recommendation is to apply two inches of water each time you irrigate, regardless of the water zone being irrigated*. Water zones requiring less irrigation will be watered less frequently than those requiring more irrigation, but when water is applied, it is best to apply enough water to ensure deep penetration. This places most of the water where it is protected from evaporation and can remain available for a longer period, and also encourages deep rooting.

Our watering recommendations are based on the principle that all the plants within a watering zone have similar water requirements. Applying the correct number of supplemental inches of water to the entire planting, even if it is made up of shrubs, trees, grasses, and wildflowers, should provide sufficient but not excessive water to all. The main distinction among different kinds of plants within a watering zone can largely be reduced to differences in plant size, that is, larger plants within a given watering zone need more water than smaller plants. These differing needs are met by supplying different sizes of plants with different numbers and types of emitters.

When designing your drip system, keep in mind that the system will need to be able to supply sufficient water to the plants after they reach mature size.

Obviously, a small sapling will not need as much water as a fully grown tree, but the initial system should allow for growth. The best way to do this is to increase the capacity of adjustable microsprinklers or to add point-source emitters as the tree develops. For point-source emitters on plants that will change greatly in size through time, it is a good idea to run a "rat-tail" line (a side line that attaches to the main drip line using a tee-connector) that is long enough to accommodate all the emitters that the mature plant will need. This way, you can add more emitters and widen the spacing of the loop as the plant grows, without needing to move lines or add more line. Make sure that you do not push the number of emitters beyond the capacity of the supply line. When you are calculating water demand on the supply line, you need to use the capacity of emitters that will water the mature tree, not the number that you plan to install for the sapling.

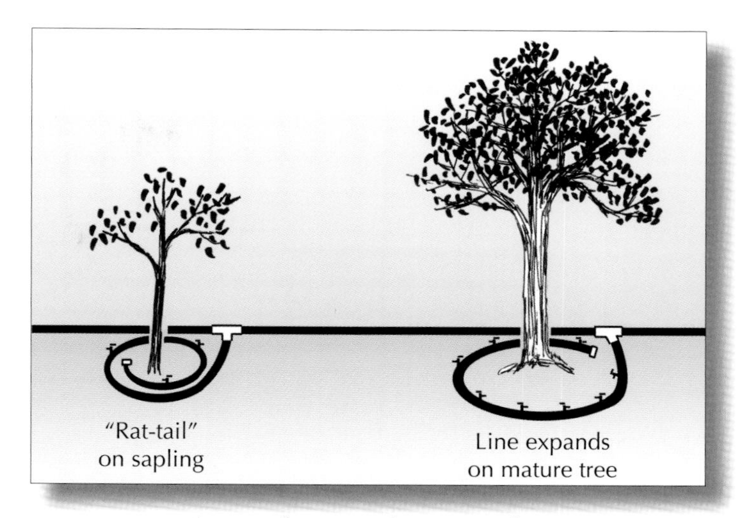

A rat-tail on a sapling expands as the tree matures.

It is usually easier and more practical to place large plants such as trees on their own supply line, separate from the supply line for smaller plants, even within a water zone. This is because the absolute amounts of water to be applied are so different that smaller plants would be overwatered, even with a minimum number of emitters at the lowest flow rates, at the longer durations needed for larger plants. In order to water more of the crown area of a tree, it is often better to use microsprinklers than to multiply the number of drip emitters excessively. Combining microsprinklers and drip emitters on the same supply line can be problematic because of pressure issues. Microsprinkers also have higher flow ratings, and when the trees are mature, they could take up most of the capacity of a single supply line. Having the trees on their own supply line also introduces the possibility of multiple water zone plantings, with the trees in an irrigated water zone but the understory and interspace plantings in a zone that does not require supplemental irrigation.

*Random Versus Regular Emitter Spacing*

The drip system we have been discussing so far is designed to target each plant individually. Such a system is called a random spacing system because the placement of the lines and emitters follows the placement of the plants rather than being arrayed in a regular grid. The underlying reason for choosing a random spacing layout is that when plants are widely and irregularly spaced, targeting water to the root zone of each plant is the most efficient way to apply it. Most of the planting areas in a native plant landscape will be

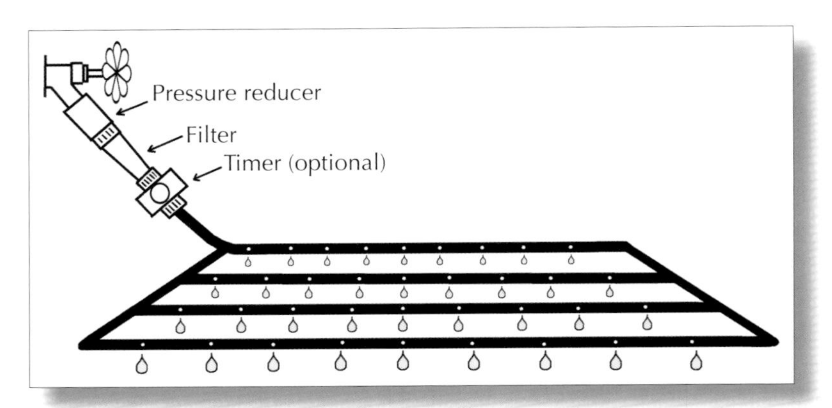


Layout of a fixed-spacing drip system.

suited to watering with a random spacing system, because the plans generally call for relatively wide spacing, mainly to avoid root zone overlap and the resulting competition for water. Wide spacing is also quite characteristic of many natural plant communities in the region. To capture the aesthetic of such communities, this spacing needs to be maintained.

The other type of irrigation layout is called grid or fixed spacing. There is definitely a place for fixed spacing drip systems in native plant landscaping. Fixed spacing drip is a practical way to irrigate a densely planted native landscape, such as a ground cover of trailing daisy or a blue grama lawn. In other words, fixed spacing drip systems can be a water-saving alternative in situations that would traditionally call for overhead sprinkler irrigation.

The simplest fixed spacing system uses inline emitters and is easy to envision, engineer, and install compared to the random spacing layout. All you have to do is lay out the inline emitter tubing in parallel rows to create a grid of emitters. The assumption with this design is that water will move laterally through capillary action to create a solid wetting front that provides uniform water. How you lay out inline drip tubing in a fixed-spacing grid will depend on your soil. As mentioned above, the pore spaces in sandy soil are large, water flows downward quickly, and there is little lateral movement, so you need to use tubing with a relatively narrow spacing between emitters, and have only a short distance between the lines. With a loamy or clay soil, lateral water movement is greater, so you can use a wider spacing between emitters and between rows. We recommend twelve-inch spacing in sandy soils and eighteen-inch spacing in loamy or clay soils for fixed spacing drip irrigation grids. This spacing creates considerable overlap in emitter coverage, and ensures an ample and uniform near-surface water supply that simulates rainfall across the entire area, especially necessary for plants like blue grama, which have relatively shallow roots.

Sometimes it will be advantageous to combine random and regular spacing in the drip system for a single planting. In this case, it may be easier to use regular drip line and installed emitters for the grid portion of the area, rather than inline emitters. For example, suppose you have an area planted to trees and shrubs that are fairly widely spaced and require targeted watering, but that these areas interfinger with others that have a ground cover of bunchgrasses and perennial flowers at closer spacing, with interspaces narrower than the average crown diameter of the plants. Targeting every one of these small plants with its own emitter would create a maze of snaking lines and would probably result in overwatering. In these areas, it would be better just to lay out lateral supply lines parallel to each other with an eighteen- to twenty-four-inch spacing and to install emitters at eighteen- to twenty-four-inch intervals, creating a subsection of the system with regular spacing. It is not important that these lines be perfectly straight—they can go around plants if necessary. The point is to optimize the number of emitters to a level that provides adequate water to all the plants without creating greatly overlapping coverage and wasting water because emitters are too close together. The emitters do not need to be as close together as in the ground cover or lawn scenario, because the bunchgrasses and perennial flowers are generally not shallow-rooted. You could use inline drip line for this purpose, but mixing systems from different manufacturers can be tricky. You need to make sure that you have the correct connectors for interfacing the two types of drip line and that they work optimally over the same pressure range.

## Designing a Drip Irrigation System

Now that you have some idea of how a drip system is put together, it is time to take out a copy of your planting plan and apply this knowledge, along with information about your site and the plants you are going to grow, to the design of your irrigation system. You will begin by sketching in a provisional watering system. For planting areas that will not need regular water, the most important consideration is the availability of water for establishment and for occasional hand or portable sprinkler irrigation during exceptionally dry periods. If you have the option of installing hose bibbs, install one or more within easy reach of all planting areas that will not have a permanent system. At least try to plan a route from an existing hose bibb, either along a path or across an unplanted area, to provide hose access to each area.

For planting areas that will have permanent drip irrigation systems, the planning can take one of several courses at this point. If you have an existing sprinkler system that you propose to convert to drip, the current location of sprinkler heads will form the basis of your design. If you plan to operate your system aboveground from one or more hose bibbs, the location of these bibbs will provide the starting point. If you plan to install an independent drip system from buried PVC pipe, the design of your system will need to include the point at which your buried system ties into the main water supply, the layout of the buried lines, and the location of risers where drip lines will be attached, as well as a plan for how the drip lines themselves will be laid out. Guidelines for designing and installing a drip system that includes buried PVC pipe are beyond the scope of this book, though the principles we describe here apply to the parts of the system that involve PE tubing and emitters.

To give you a clearer idea of the watering system planning process, we provide a hands-on example for the landscape illustrated in our design sequence. The site description tells us that this landscape is situated in the semi-desert water zone, with a mean annual precipitation of fifteen inches. Areas on the landscape plan that are designated as minimal (desert) and low (semi-desert) water zones should not require a formal irrigation system, especially as this site is at the high end of the semi-desert zone. The principal areas that will need drip irrigation are the three medium-water-zone canopy plantings and the high-water-zone plantings close to the north and east walls of the house. The understory plant species in the medium-water-zone canopy plantings are comprised entirely of bunchgrasses and perennials from the low water zone. It would be OK to provide these with supplemental water along with the trees and shrubs—but this really is not necessary for low-water-use plants. Because drip irrigation can target the individual trees and shrubs, we can in effect have two water zones in the same general area. This greatly simplifies the job of irrigating these areas. To avoid overwatering the low-water-use shrubs in this area, namely the cliffrose, they too can simply be excluded from the drip system.

Three hose bibbs already available on the outside of the house will provide the water for the irrigation system. These have three-quarter-inch supply pipes, so the maximum number of gallons per hour on each line should not exceed two hundred. Divide the yard into areas that will tentatively be supplied with individual lines, and determine which hose bibb will supply each area. For example, the medium-water mid-height and canopy areas in the back yard make a logical grouping to water on a single line from the back hose bibb. Use an irrigation design worksheet like the one shown here to determine whether your supply pipe can provide the plants you plan to water on each single line with sufficient water to meet their needs when mature, and determine how long you will need to water each time you irrigate to apply the requisite two inches of irrigation for each line. Once you have determined that the line as designed will supply the mature plants and will not have a run that exceeds two hundred feet, you can calculate the water demand and the length of an irrigation event for the new planting. Then you can determine what your emitter needs will be for each plant.

Specifying emitters for the near-term scenario is most important now, but it is also a good idea to consider how you will eventually water the mature plants. There are many correct ways to specify emitters for a given number of gallons per hour, and each has its pros and cons. In general, emitters with higher flow rates tend to water more deeply, while increased emitter numbers spread the water more widely within the crown area. Once the trees begin to approach full size, you will probably want to switch to microsprinklers, which can cover more crown area with a given volume of water. If you use the "rat-tail" method

## Designing the Watering System

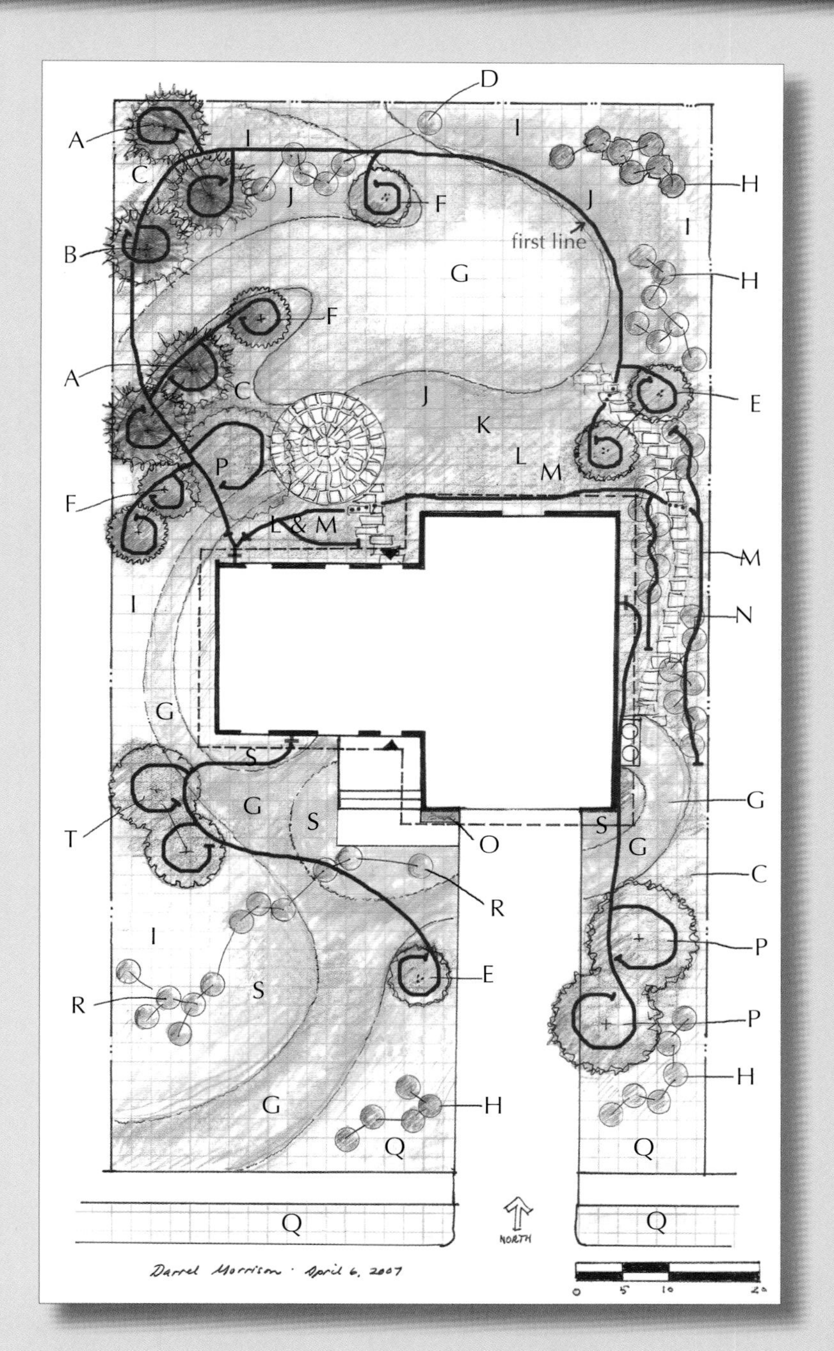


The goal for watering is to design an efficient and functional drip system that includes separate lines for plants in different water zones and for plants within a water zone that differ greatly in size. For each line, make sure not to exceed the maximum number of gallons per hour for the supply pipe, and ensure that the maximum run of line is less than 200 feet. This landscape design is efficiently watered with four lines, three for the medium-water-zone plantings and one for the high-water-zone planting. The low- and minimal-water-zone plantings do not need a permanent watering system.

1. The first line starts at the back hose bibb and waters the trees and shrubs in the back yard medium water zone. Note that the understory plantings are low-water-zone plants that do not require extra water.
2. The second line also starts at the back hose bibb. It waters the high-water-zone plantings near the north wall and on the east side of the house.
3. The third line starts at the hose bibb on the east side of the house. It waters the medium-water canopy planting to the east of the front driveway.
4. The fourth line starts at the front hose bibb. It waters the canopy plantings on the west side of the front driveway.
5. The hose bibbs can also be used to provide occasional extra water to the low-water plantings during exceptionally dry times.
6. The minimal-water plantings should not need extra water after establishment, even during dry times.
7. Remember, all plants need extra water during establishment.

## Watering by the Numbers

### *6. Making a Drip Irrigation Design Worksheet*

Irrigation at Maturity—8 hours per irrigation event
*drp = drip emitter; ms = microsprinkler*

| | Plant no. | Crown diam. (feet) | Gallons per plant for 2″ | Total gallons planned | GPH per plant | Emitters per plant in the example | Total GPH | Total gallons delivered |
|---|---|---|---|---|---|---|---|---|
| Rocky Mtn. Juniper | 4 | 12 | 144 | 576 | 18 | 2 (8 gph) ms | 64 | 512 |
| Pinyon Pine | 1 | 15 | 225 | 225 | 28 | 2 (15 gph) ms | 30 | 240 |
| Bigtooth Maple | 1 | 25 | 625 | 625 | 78 | 5 (15 gph) ms | 75 | 600 |
| Utah Serviceberry | 3 | 6 | 36 | 108 | 4.5 | 1 (5 gph) ms | 15 | 120 |
| Alderleaf Mtn. Mahogany | 3 | 5 | 25 | 75 | 3 | 1 (3 gph) ms | 9 | 72 |
| Total Planned | | | | 1609 | | | 193 | 1544 |
| **Irrigation First Year—2 hours per irrigation event** | | | | | | | | |
| Rocky Mtn. Juniper | 4 | 3 | 9 | 36 | 4.5 | 2 (2 gph) drp | 16 | 32 |
| Pinyon Pine | 1 | 2 | 4 | 4 | 2 | 1 (2 gph) drp | 2 | 4 |
| Bigtooth Maple | 1 | 4 | 16 | 16 | 8 | 4 (2 gph) drp | 8 | 16 |
| Utah Serviceberry | 3 | 2 | 4 | 12 | 2 | 1 (2 gph) drp | 6 | 12 |
| Alderleaf Mtn. Mahogany | 3 | 1 | 1 | 3 | 0.5 | 1 (1/2 gph) drp | 1.5 | 3 |
| Total Planned | | | | 71 | | | 33.5 | 67 |

To calculate water demand and emitter needs for a proposed drip supply line:

1. Make a list of the plants on the proposed line, listing species and the numbers of each.
2. Note the crown diameter at maturity for each species (from the Plant Palette table).
3. Calculate the gallons per plant for a 2″ irrigation (crown diameter squared).
4. Calculate the total gallons planned by multiplying plant number times gallons per plant for each species and summing these numbers for all species.
5. Divide the total gallons planned by the gallons per hour (gph) that your supply line can deliver to get an estimate of the number of hours you will need to irrigate when the plants have reached maturity in order to apply 2″ of water: *1609 gallons divided by 200 gallons per hour = approximately 8 hours.*
6. Calculate gph for each plant by dividing the gallons needed for a 2″ irrigation by the hours of irrigation: For pinyon pine: *225 gallons divided by 8 hours = 28 gph.*
7. Determine a combination of emitter capacity and number that yields approximately the appropriate number of gallons: *two 15–gph* microsprinklers = 30 gph.

Repeat the calculations using the crown diameter at planting for each plant to determine the water demand for the new planting. This will usually be much less than for the mature planting, especially for woody plants that increase greatly in size.

we described earlier, with spur lines to each plant, you can add capacity to your system every year or two to accommodate the growth of the shrubs and trees without undue trouble. It usually is not necessary to change out or add emitters for perennials, which rarely have a crown area larger than a single emitter can supply. You may sometimes find that you cannot easily come up with an emitter combination that gives you the exact number of gallons per hour specified. But as long as you are ballpark close—say, within 10 percent or so—the system should work well, and your plants should get enough water.

You can approach design of any drip system using a strategy similar to the one just described. It is helpful to place Y-connectors at each hose bibb, so that you can easily attach a garden hose for hand watering or any other water use that arises without disconnecting the drip system. This is especially important for the back hose bibb in this design, in light of the fact that you may want to provide occasional supplemental water in the summer to the blue grama lawn area adjacent to the patio, especially if you live in an area with little natural summer precipitation. This grass can survive perfectly well without much summer rain. But if you want a lawn that can be mown for a children's play area, as indicated in the human needs analysis, it will perform better if it receives an inch or two of supplemental water every few weeks. A portable sprinkler could easily accomplish this task. It won't be necessary to water the blue grama on the west side of the house, as your expectations for this area would probably not be as high.

## Emitter Needs for Watering Plants of Different Sizes with Drip Irrigation

The goal is to apply the approximate number of gallons needed to add the equivalent of 2″ of water over the entire crown area of the plant.

The total number of gallons needed is obtained by multiplying the crown area of the plant in square feet by 1.25, a conversion factor which represents the number of gallons needed to apply 2″ of water to one square foot.

| Plant Crown Diameter (feet) | Crown Area (square feet) | Gallons for a 2″ irrigation | Example Drip Emitter Combinations for Applying ca. 2″ of Water during Irrigation Periods of Different Durations | | |
|---|---|---|---|---|---|
| | | | 2 hours | 4 hours | 8 hours |
| 1 | 1 | 1 | 1-1 1/2gph | --** | --** |
| 2 | 3 | 4 | 2-1gph | 1-1gph | 1-1/2gph |
| 3 | 7 | 9 | 4-1gph | 2-1gph | 2-1/2gph |
| 4 | 12 | 16 | 4-2gph | 2-2gph | 2-1gph |
| 5 | 20 | 25 | 6-2gph | 3-2gph | 3-1gph |
| 6 | 30 | 36 | 9-2gph | 9-1gph | 5-1gph |
| 7 | 40 | 49 | 6-4gph | 6-2gph | 6-1gph |
| 8 | 50 | 64 | 8-4gph | 8-2gph | 8-1gph |
| 9 | 65 | 81 | 10-4gph | 10-2gph | 10-1gph |
| 10 | 80 | 100 | 13-4gph | 13-2gph | 13-1gph |
| 11 | 95 | 121 | 15-4gph | 12-2gph | 15-1gph |
| 12 | 110 | 144 | 18-4gph | 18-2gph | 15-1gph |
| 13 | 135 | 169 | --* | 11-4gph | 18-1gph |
| 14 | 155 | 192 | --* | 12-4gph | 11-2gph |
| 15 | 180 | 225 | --* | 14-4gph | 14-2gph |
| 16 | 200 | 256 | --* | 16-4gph | 16-2gph |
| 20 | 310 | 400 | --* | --* | 13-4gph |
| 25 | 500 | 625 | --* | --* | 20-4gph |

Rule of thumb: Gallons needed for a 2″ irrigation = crown diameter$^2$ ( = crown diameter x itself).

* Better watered with microsprinklers.
** Irrigation period too long for small plants - smallest emitter overwaters.

*Mountain ash*

# How to Install Native Landscapes

Chapter Four

## Planning the Installation

Throughout the process of designing your native landscape, it has been necessary to keep referring back to the realities of your site. Now it is time to go outside and make those changes that need to be made in order to prepare your site for its new inhabitants, and then to plant them in a way that guarantees that they will prosper. This will require planning. Much of this process will probably be familiar to you from other landscaping and gardening projects you have undertaken, but there are some things that are unique about native plant landscaping, and these require careful attention. Just how complex the process will be depends on several factors. First, you probably need to deal with removing at least some of the existing vegetation on your site, whether it is lawn, foundation plantings, new weeds that inevitably show up uninvited in recently-spread topsoil, or longstanding infestations of perennial weeds. Second, depending on the nature of your soil, you may need to do some soil replacement or terracing/berming, or both, to create the drainage that your plants will need. These kinds of modifications may also be necessary even if you do not have drainage issues, for example, if you are trying to create a congenial place for plants from much drier water zones. And, as discussed in the design section, terracing or berming can also be used to create topographic relief solely for design purposes, not specifically to meet the cultural requirements of plants. You may also need to make some grade modifications in order to implement the water harvesting system you have designed. These two steps can be relatively simple or quite complex, depending on the magnitude of the changes you need to make.

When these major modifications are more or less complete, the next step is installing, or at least roughing out, the hardscape—paths, patios, and shade structures, for example. This step should be carried out more or less simultaneously with laying out the main lines or drains for any irrigation or water harvesting system you plan to install. This is especially true if installing the system involves trenching. For example, you will want any buried pipes or French drains in place under the path system before finishing the paths. The hardscape also includes any large rocks you plan to place in the landscape.

Artful use of landscape rock at the LDS Conference Center in Salt Lake City.

When the hardscape and irrigation system have been at least provisionally dealt with, you are ready to groom the planting areas to receive plants. This involves tilling in any amendments such as compost, sand, or gravel, and perhaps raking to shape the beds. If you are installing a drip irrigation system, the drip lines that will actually deliver water to the plants (as opposed to the main supply lines, which may intersect hardscape) will be laid out at this point, after the beds are shaped. The next steps are to obtain the plants, lay out the planting plan on the ground, and finally, to plant.

When the plants are safely planted and watered in, you will spread any needed mulches, then sit back and drink a well-deserved lemonade. Actually, you will get lots of well-deserved lemonade breaks, because obviously, for even a modest project, the installation process may take several days. More complex projects may take weeks or possibly even months. Do not be discouraged by this—your patience, planning, and hard work will lead to a beautiful, interesting, low-maintenance native landscape that will reward you for your initial efforts for many years to come.

You know where you are in the process of actually creating your native landscape—at the very beginning. And you know where you want to be—enjoying that lemonade in the shade, looking out over your thriving native garden, satisfied with your work. To make this journey as painless as possible, the key is the development of a realistic timeline for completion of the project. It is often less daunting to tackle a large project in phases, so that the expenditures of both energy and money can be spread out over time, yet completion of the first phase can quickly begin providing satisfaction. This point, mentioned in the design section, is reiterated here, because the timelines for different phases of a multi-phase project will be different from each other. Be sure to read through the entire section on installation before you develop your installation timeline, so that you can get some idea of the magnitude of various kinds of tasks.

To begin to develop a timeline, pick a provisional planting time for your project, or for the first phase of a multi-phase project. Spring may seem the obvious time to plant, as that is the traditional timing for frost-sensitive vegetables and flowers. However, fall is also a very good time to plant, particularly for a native landscape. With a spring planting, the plants are still "rooting in" (extending their roots out into the soil) when summer arrives, with hot weather and water stress. Fall planting works with nature because your plants grow roots when the soil is still warm but when air temperatures are cooler. Many native plants continue to grow roots through the winter, especially under the snow, so that when spring arrives, they are well rooted in and ready to tolerate the hot, dry weather to come. Spring plantings can be successful, but frequent and careful watering to establish them is much more important than in fall plantings, which can often rely largely on natural precipitation for establishment.

Planting in the fall does have one practical drawback, at least as things currently stand. Because people are so fixated on traditional spring planting, most plants that are sold, including native plants, are produced for sale in the spring, making it difficult to find plants for a fall planting. Hopefully this will change, as the market for native plants increases. In fact, this could be a good way for nurseries to diversify and increase their fall sales.

The next step in preparing a timeline is to make a list of the activities you need to carry out. Sort the list into the order in which the activities need to be completed and make an estimate of the time required for each activity. This will depend on the budget available in some cases. Hiring someone to do a job is generally faster than doing it yourself, although you are often at the mercy of someone else's scheduling for contracted work. Similarly, purchasing rocks and having them delivered is faster and easier than picking them up yourself at a public quarry, but it is also much more expensive. There is a limit, of course, to the size of rocks that can be safely handled without the specialized equipment used by professionals.

Some tasks can only be performed at certain times of the year, and this will need to be factored into the timeline. For example, any work that involves heavy equipment must be performed when the ground is dry. Driving heavy equipment over wet soil causes serious compaction that can be very difficult to eliminate or manage. In fact, compaction caused during construction is one of the main obstacles to successful landscaping around newly constructed houses. It can take decades for this compaction to abate through natural processes like freezing and thawing. Mitigate this problem by keeping heavy equipment off your site when the ground is wet.

# Carrying Out Site Modifications

## Taking Out Existing Plantings

Removing a lawn by solarization.

Getting rid of existing lawn is often the single most problematic aspect of site modification for people retrofitting an old landscape to make it suitable for native plants. Make no mistake, Kentucky bluegrass can be a pernicious perennial weed in a landscape where it is not wanted. It is important to do a good job on removal, or this plant can come back to haunt you and greatly increase your maintenance chores.

There are basically three methods for taking out a lawn: solarization, herbicide, and sod-cutting. The fourth possibility, hand removal, with a mattock, for example, is so backbreaking that it cannot realistically be completed in an acceptable timeframe, and it is not as effective as other methods at completely eliminating a lawn. Solarization involves soaking the lawn down well, then covering it with a transparent plastic sheet that is buried along the edges. The idea is that heat will be trapped underneath the plastic and will kill the lawn. This method works only during the hottest, sunniest time of the year and takes four to six weeks. It works best on flat surfaces. Effective solarization can also kill weed seeds in the surface soil, which is an added advantage to this method. It may also kill disease organisms that could potentially harm your native plants. Clear UV-stabilized plastic, one to three mils thick, works best; UV stabilization keeps the plastic from deteriorating before its job is done. If the plastic is too thick, it is hard to get the tight fit over the ground that is necessary for effective heating.

Taking out a lawn with herbicide is quite a simple process. Glyphosate, which has a relatively low environmental impact, is a suitable choice. The lawn must be actively growing when it is sprayed, but this can be any time from early spring through late fall. It will take longer to see the result if the spraying is carried out in cool weather, but the herbicide will be just as effective. Spray in fair weather, as rain can wash off the herbicide before it has a chance to penetrate. It must penetrate through the leaves; it loses its effectiveness once it contacts the soil. Be sure to choose a calm day, so that there is no chance of drift or overspray onto nontarget plants in your yard or your neighbor's yard. Glyphosate is non-selective, and even a little overspray onto other leaves can cause damage or death, especially to sensitive plants. And, just for the record, most native perennials are very sensitive to this herbicide. One disadvantage to taking out lawn with herbicide is that all the weed seeds are left behind, ready to sprout up in your planting at some inopportune moment in the future.

Sod-cutting is probably the most labor-intensive method of lawn removal, after hand removal. The appropriate piece of power equipment can be rented for a few hours for a nominal sum, and even a small adult can manage a sod-cutting machine. The real work is dealing with the ribbon-like strips of sod that are left behind. These can be cut in manageable sections and rolled for easy handling, but even so, they are heavy. Then there is the question of what to do with all that sod. If you are a vegetable gardener, a use will immediately suggest itself. Composted bluegrass sod is a great soil amendment for vegetable gardens. And, if you can arrange for the composting process to be "hot," weed seeds will be killed along with grass rhizomes. Composting the sods is probably the best idea even for non-vegetable gardeners. Once the grass and weed seeds are dead, the topsoil can be reclaimed for other uses or given away—or perhaps a vegetable gardener friend can take the sods for composting.

Advantages to the sod-cutting solution include the simultaneous removal of most weed seeds as well as the reduction in organic matter and fertility that results when the top two inches of soil are removed. If you solarize or spray your lawn, the rotting sod that remains in place will greatly increase organic matter and fertility. This is true even if you rake off the sods once they have begun to decompose. As explained earlier, high organic matter and fertility present little or no advantage for most natives and may actually be detrimental to some. Disease problems may be much reduced with solarizing, but this is largely an untested idea in the context of natives. If you do opt for sod-cutting, be sure to set the blade at least two inches deep. Otherwise, deep-seated grass rhizomes may be left behind.

Most of the other plants that you will need to remove to make room for your native landscape will be easier than turf. Herbaceous plants can just be pulled or dug, after wetting the ground to make sure you get the complete root system, especially for root propagators like irises or daylilies. Small shrubs can be clipped off at ground level, then dug or pried out with a spading fork or shovel. Larger shrubs like junipers may require the help of a machine for removal. Chaining the juniper trunk to the sturdy steel bumper of a pickup truck, then driving the truck slowly away, is a method we have used successfully—it helps if the soil is wet. Cutting off a shrub or tree flush with the ground may be sufficient for many purposes, as long as you are not dealing with a species that resprouts readily. For larger trees, it may be best to call in the professionals.

## Weed Control

The importance of a good program for weed control prior to planting your native landscape cannot be overemphasized. Weeds are the reason most often cited for the failure of native landscapes. Part of the reason for this is that people have the idea that native landscapes are somehow less vulnerable to weeds than traditional landscapes. This is not true. In the weed-infested, highly disturbed environment where most humans live, all landscapes are vulnerable to weed invasion. It is true that there are ways to manage native landscapes to minimize weeds, but the very first management tool to apply is to eliminate as much of the weed problem as possible before planting. Your site analysis has provided information on which weeds will likely cause problems for your landscape. Different types of weeds require different control approaches.

If you have a serious problem with annual weeds, a method called fallowing can help. This involves watering to encourage a crop of seedlings, killing the seedlings using mechanical means (hoeing, tilling) or herbicides, then watering again to bring up another crop. If this process is started in the spring and repeated throughout the summer, much of the bank of annual weed seeds can be depleted, especially if the surface is mechanically disturbed each time. Both spring and summer annuals can be controlled using this method. If you know your problem is a summer annual such as pigweed or purslane, you can start when the weather warms up, usually in late May or June. If winter annuals like cheatgrass are your problem, the elimination of growing plants in late spring, followed by a fall fallowing program, can usually go a long way toward control. If you aren't sure what assortment of annual weeds you have, just starting in spring and carrying on through the fall will cover the bases. Fallowing will not completely eliminate the seed bank, but, combined with targeted watering and mulching after planting, it will reduce the chore of hand weeding considerably, without the need for herbicides to control annual weeds.

If your problem is persistent perennial weeds, you may have a more difficult control task on your hands. Hand pulling is generally an effective control method only if you are diligent over an extended time period, a period measured in years. Some of these weeds, such as whitetop and quackgrass, can be effectively killed with relatively benign herbicides such as the glyphosate used to kill Kentucky bluegrass. Others, such as the notorious field bindweed, aka wild morning glory, are difficult to kill even with powerful herbicides. The reason this plant is so hard to kill is its very deep and very extensive root system. Many recommend spraying in the fall with a cocktail of equal parts glyphosate and another herbicide called 2,4-D, when the plants are actively translocating from the leaves to the roots in preparation for winter dormancy.

Bindweed emerging through an asphalt walkway.

A serious field bindweed infestation may best be dealt with by turning the area into a patio, though even then this weed will do its best to escape along the edges, or even rear up through cracks in the cement. Alternatively, if the area is planted to widely spaced shrubs, well-pinned professional-grade landscape fabric can be used under the mulch in the interspaces to suppress bindweed. It will still try to come up in the bare areas around the shrubs, especially if they are irrigated, but at least the area to weed will be smaller. Solarizing before planting is another option that will at least set bindweed back and possibly eliminate it if circumstances are optimal. Covering an area completely with landscape fabric and mulch will eventually kill bindweed, but this may take almost as long as constant hand pulling. It is, however, a lot less work. Bindweed seeds live a long time in the soil and survive all control treatments except solarization. Constant vigilance is required to prevent reinfestation from seedlings, as the plants rapidly get too tough to eliminate by simple hand pulling. But if you can pull out each seedling by its taproot, the plant will truly be dead.

Weeds will be a fact of life in a native landscape, as in any other landscape, but you can reduce maintenance chores later by spending the time and attention to deal with the weeds as effectively as possible before you plant. One other tip to prevent a major step backward after all your hard work on weed control is to be sure not to introduce any soil or soil amendment into your planting site unless you know for certain that it does not contain weed seeds, particularly the seeds of perennial weeds.

## Dealing with Drainage

If you are among the unlucky who determined during site analysis that your site has serious issues with drainage, then this section is especially for you. First, take heart, there is a way out of this dilemma. If your drainage problem is surface compaction caused by heavy equipment, probably during construction, then a process called "ripping" may go a long way toward a solution. Ripping means gouging long furrows into the surface to break up the compaction. This is usually done with a chisel implement that is attached to a farm tractor, backhoe, or small caterpillar tractor. The first consideration when hiring someone to rip your compacted soil is to make sure that it happens when the soil is dry. Otherwise the heavy machine will cause more compaction than it removes. Caterpillar tractors are track vehicles that cause less compaction than regular tractors, but they are harder to find. If you are not up for the heavy machinery approach, another way to rip a small area is to use a power tree auger to punch numerous holes through the compacted layer. The holes will permit water penetration and will greatly increase the rate at which compaction progressively becomes less severe.

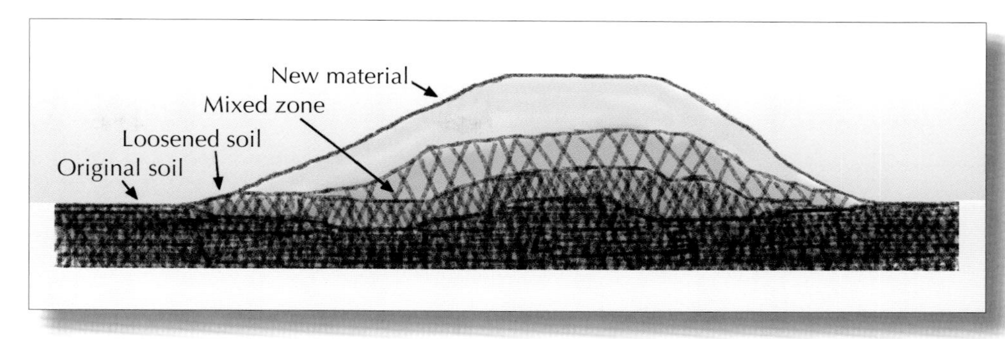

Inner structure of a berm.

If your problem isn't compaction per se, but a heavy soil that drains very slowly, a salty soil, or a water table very near the surface, you need to try other tactics. Often all three of these problems occur together in low-lying valley areas. If your heavy soil is a clay subsoil on a more upland site, you will have an easier time, because your problem will probably not be compounded by salt or water near the surface. The basic idea for solving poor drainage that is not caused by compaction is to elevate the planting area above the level of the problem, that is, to help the plants "escape" the excess water and salt. This is accomplished by terracing or berming, often with coarser-textured soil materials brought from offsite. It could involve building retaining walls that are then backfilled with the coarse materials, or berming or mounding onto level ground. Which approach you use will likely be dictated by design considerations. If your entire property is affected by poor drainage, it will probably not be practical to bury your problem soil over the whole area. Instead, you will need to pick high-visibility or pivotal areas of the landscape to modify with berms or terraces.

When layering two soil materials that have very different physical properties, it is important to blur the boundary between the layers so that roots will not stop at the boundary, creating a "bathtub" effect. First, the onsite soil should be harrowed or tilled to roughen its surface, or ripped if it is compacted. Then, six to twelve inches of the offsite material should be tilled in. The rest of the offsite material is then added on top of this mixed layer. How deep should the added layer be? The deeper the better, from the point of view of the plants. Most natives will eventually root into the underlying material, unless the added layer is very deep. But if they can keep a substantial part of their root system above the heavy layer, they will probably do better. Two to three feet is a reasonable depth to aim for, but even a foot of coarse material can be of considerable help, especially for perennials.

What should the added material be? This is somewhat a matter of opinion. Most importantly, it should not contain significant amounts of clay, as this will just compound the problem. It could be topsoil, but we do not recommend this, for several reasons. First, it would cost a lot of money to apply a two- to three-foot layer of purchased topsoil over any significant area. Second, the topsoil would almost certainly contain weed seeds. Topsoil for sale usually comes from agricultural land that is being developed. This topsoil will be infested with weed seeds, possibly the worst kind of weed seeds, whether or not the seller tells you it is weed-free. No topsoil is weed-free, and there is little enforcement of regulations to control the sale even of topsoil infested with noxious weeds like field bindweed. Third, the topsoil would likely be high in organic matter and in fertility, two features that are not necessary for native plant landscapes, as previously discussed. And fourth, materials sold as "topsoil" can have widely varying physical properties. If you purchased a clay loam topsoil, this would add significant amounts of clay. Thus in most cases it would be a waste of money or worse to use topsoil to create an "escape" layer for your plants. Another material that can be used is sand or mixed sand and gravel. This can be purchased from local quarries and will have the advantage of coming from a recently-made hole deep in the ground. It will be fast-draining, low in organic matter and fertility, and weed-free. We believe that this is the best material to use in raised beds built to alleviate poor drainage. If the plants you intend to use need more organic matter, fertility, and water-holding capacity than sand can provide, it is a simple matter to till weed-free organic amendments into the surface as part of the bed preparation. What is surprising is how many native plants can thrive in unamended sand, especially if the sand contains particles of different sizes made from a variety of minerals. Over a fairly short time, the top layer of the sand will acquire a more organic character, from leaf fall, root turnover, and sediments trapped from the wind. If you mulch the surface, this will happen more quickly, especially if you use an organic mulch. How you manage this added soil layer will depend on the water zone at your site and the water zone you are trying to create. If you make your raised bed entirely from sand and want to keep the organic matter low, it helps to mix in coarse gravel in the top few inches to stabilize the surface. Otherwise it will shift around every time you walk on it, which can be hazardous to the young plants. Gravel in the surface layer also makes the bed less attractive to passing felines.

If you have ordinary garden soil that does not present serious drainage problems, you will probably be able to create water zones equivalent to or wetter than the water zone at your site without creating raised planting areas or adding soil material. But you may still need to create a place with better-drained, less organic soil to accommodate plants from drier water zones or plants from your water zone that are particularly susceptible to the problems associated with high organic matter. One approach to this is exactly the same as the approach described above. Berms or terraces built from coarse soil materials low in organic matter will provide the environment that will help these plants to thrive.

Another, less drastic approach is just to amend your surface soil with coarse soil material. This would involve carrying out the first steps described above, namely ripping, tilling, or harrowing the original soil surface, adding six to twelve inches of coarse material, and tilling it in to mix the two materials. Use enough added sand and gravel to constitute at least 50 percent of the mixed layer, or you risk the creation of adobe brick rather than a friendly medium for plant growth.

Drainage issues are also involved in the installation of water-harvesting systems, but in this case, you want to direct runoff onto the planting areas. You need to install French drains and construct catchment swales at the same time that you are performing site modifications such as berming and constructing retaining walls. This is to make sure that the drains are installed at grades that permit the water to flow downhill to the target area.

Just three of the many ways to make a wall.

## Installing Hardscape

There are many fine resources dedicated to the subject of constructing paths, decks, patios, walls, arbors, and other features and structures that may be called for in your landscape design. We will not be spending much time on this subject here. We provide a list of some of our favorite resources in the section on getting more information.

People often ask what hardscape materials "go best" with native plant landscapes. Clearly, the design may be very formal or quite informal, and different materials may be suitable for these varying design styles. Wood always looks good, but it is expensive and can require considerable maintenance over the long run. Plastic-based materials made to look like wood may be a good solution for decks, arbors, and trellises. These are often made from recycled plastic, which is an added advantage. These materials are

Two examples of stone paths.

slowly becoming more available at mainstream home improvement centers, and can readily be ordered from Internet suppliers.

Retaining walls can be made of wood (recycled railroad ties, for example), plastic-based wood look-alike materials, rock, or concrete blocks. Rock is generally the high-end solution, unless you quarry it yourself and haul it home. And building rock walls is an art unto itself, as well as being incredibly labor-intensive. But concrete blocks are now available in a wonderful assortment of shapes, colors and finishes, some of them quite compellingly rock-like, and they are a lot easier than rock to stack into a wall that is plumb. Another option is to face a concrete block wall with stone.

Paths are often the most difficult hardscape features to specify in terms of materials. Based on our experience, the number one consideration from a maintenance standpoint is that the path be weed-proof. For this reason, we do not recommend bark chips as a path material, even if they are installed over high quality landscape fabric. As they disintegrate over time, they generate a fine material that closely resembles very black soil and is a magnet for weeds. These do not grow up through the landscape fabric, but root through it from the top down. Hand weeding mulched paths is a time-consuming activity that will never be high on your list, so avoid the necessity if at all possible. Gravel paths are somewhat better than bark mulch paths, as long as they are installed with landscape fabric underneath. Gravel does not fall apart to generate organic matter, but in an area with a lot of leaf fall and other windblown sources of organic debris, even gravel paths will eventually host weeds. It is better to use coarse gravel. A size called "drainpipe gravel," which includes rocks that average an inch or so across, works well. Avoid pea gravel for paths or any other landscape use. It has a way of spreading itself messily from one end of your property to the other.

The rock on the right looks like it just landed there from someplace far away. To avoid this just-landed look, prepare a landing place before the rock is delivered, so that it sits with its base below ground.

Beaten earth paths are probably as good as either bark mulch or gravel paths in terms of maintenance. By deliberately compacting the soil, you can make an adobe-like surface that is fairly resistant to weed invasion, especially if your soil is high in clay and you use the path frequently. This is the way paths form spontaneously through repeated human use. Another approach to the beaten earth path is to mix the surface soil material with Portland cement or a liquid resin, which then hardens into a weatherproof and weed-proof surface that matches the surrounding soil. The resin approach is often used for trails in national parks because of its low visual impact.

Paving stones are another option for paths, and stone paths can be very beautiful. Unless you plan to plant between the paving stones with a ground cover like mat penstemon, it is best to install landscape fabric beneath the stones to cut down on weeds. Any cracks between the stones can be filled by sweeping in sand, or the stones can be laid on a bed of sand to help with leveling. Bricks, or concrete blocks designed to imitate bricks or stone, can also make durable, low maintenance paths. Again, install landscape fabric underneath first, or else weeds will be squeezing through the cracks, no matter how tight.

Hardscape installation can be a do-it-yourself activity or it can be a job for a contractor, depending on the scale and complexity of the project and the materials used. Be sure to refer to your site analysis for the location of any underground utility lines on your property prior to beginning any major earth-moving, or even any digging or trenching. Have them marked again if you are not sure of their exact location. Again, if you hire a contractor who will use heavy machinery as part of the installation, make sure that the work is performed when the soil is dry. And, if your soil is at all heavy, provide scrap plywood for the equipment to drive over, or ask that the contractor provide it. This will further help to avoid compaction, which can be a problem on heavy soils even when they are quite dry.

Smaller landscape rocks can be carried in a wheelbarrow and placed by hand with a shovel or prybar, but large rocks must be brought in and positioned using heavy equipment. Larger rocks, especially, need to be partially buried to create a more natural appearance. Expensive rocks are often dumped down willy-nilly on the surface, especially in commercial landscapes. To place a rock properly, you need to find the location where it will be installed and excavate a bed to receive it, before it is delivered. If the rock is of any size, you will not be able to do this after the rock is in place. If you know the dimensions of the rock and how far into the ground you want it to sit, you can make sure it is placed the way you intended.

## Installing the Watering System

The process of installing a drip system is a stepwise process that extends from the time you are performing site modifications and installing hardscape until after you have actually planted. You need to have a good idea upfront of the lay-out of the watering system you intend to install, because supply lines may need to intersect hardscape features such as paths, and the best way to do this should be decided before either the hardscape or a water line is installed. You will prob-ably want to get a cost estimate of the materials you will need: PE tubing, fittings such as filters, pressure regulators, valves, connectors, and end caps, as well as the emitters themselves. This involves estimating how much line you will need, how many fittings of various types, and how many emitters of each type. Once you have finalized your design, you can make a reasonably accurate list of all the parts you will need.

When you go to purchase the materials for your drip system, it is often a good idea to bump up your estimates for line, connectors, and emitters by five per-cent to avoid repeated trips to the irrigation supply store during installation—the installation process may not be as precise as the plan makes it look, and it may not go quite as smoothly. Any materials you don't use right away will come in handy later as you perform maintenance on the system.

Remember to include tools such as a punch tool for installing emitters on your list of needed items. Once you punch a hole for the emitter, it is a simple matter to pop it in by hand. A sharp pair of bypass pruners is a good tool for cutting PE tubing, and a sharp knife also works well. Extra connectors are useful for repair-ing accidental tubing cuts during or after installation. To repair tubing, you just cut out the short piece that is damaged and attach the ends together with a con-nector. Both regular PE supply line and the inline emitter line can be repaired in this way, as long as you use a coupler that matches the diameter of the line. It is also good to have a supply of the "goof plugs" we mentioned earlier on hand—these are stoppers used to plug holes where emitters are no longer wanted, or where there is a leak around the base of the emitter. To replace an emitter found to be damaged or defective, just remove the emitter with a pair of pliers and install a new one. To repair a leak, remove the old emitter, install a goof plug in its place using gentle pressure, and re-install the emitter a short distance down the line. Sometimes the goof plug will need to be sealed with a waterproof sealant to totally eliminate the leak.

When first attempting to insert PE line into a compression fitting, you may become convinced that it is impossible. Try dipping the end of the line into very hot water right before you try to push it into the fitting. Another solution is to use some type of lubricant on the outside of the line. If the line really, really won't go into the fitting, check to make sure that the fitting size actually matches the tube diameter. Cramming 18 mm tubing into a 16 mm compression fitting truly is impossible.

Another essential for installation is metal staples for pinning down the drip line to prevent it from wandering or creating tripping hazards. Irrigation supply companies sell steel staples made for this purpose, but the staples made for holding down jute erosion fabric are cheaper. These tend to rust in place, which makes them stay put better than steel staples. By the time they have rusted out completely, the line is usually pretty well trained to stay in place. Homemade staples made of stiff wire, e.g., from old surveyor flags or coat hangers, also rust in place and work quite well.

Whether installing a system from a hose bibb, from risers on an underground system, or as a retrofit to an existing system, you need to start at the point of supply and work outward, laying out the main supply lines first. If possible, it is best to keep all the valves in a box close to the water supply point, and to run the mains for each of the subsystems out from this valve box. This is especially important if the valves are to be electronic and wired to a timer, because it shortens the wiring runs, greatly decreasing the chance of problems. If the system is a retrofitted automatic sprinkler system, these valves will probably already be in place. If the valves are manual, it is not so important that they be close to the water supply point—ball valves where lines branch can be used to switch over the flow of water from one subsystem to another.

Once the main lines and valves are in place, it is time to lay out the supply lines that will actually provide water to your plants. While it is imperative to have a plan for your drip system in hand at the early stages of planning the installation, surface supply lines should be laid after the site is graded and the planting areas are in their finished configuration.

It is easy to over-engineer your system if you try to figure every last thing out before you start laying out the lines—and the plants will probably not be placed with the military precision suggested by the planting plan. So, never install emitters into the drip line in a random spacing system until the line is actually in place next to the plant to be watered. For a fixed spacing system, it can be easier to install the system before planting, then place your transplants in the grid interspaces.

A drip system becomes much less noticeable as the plants mature. These are "before" and "after" pictures of the same area.

To make the drip line easier to handle and to get it to stay uncoiled, it helps to let the rolls warm up first in the sunshine. But it is better to adjust and stake the line and to install the emitters when temperatures are cool, because drip line expands in the hot afternoon sun. If you install the emitters when the line is hot, they can change position when cold water runs into the line, especially if the line is long. Once the lines and emitters are in place they can be mulched over, but drip lines and emitters should not be buried in soil. Burrowing animals love to destroy PE tubing, and it is difficult to troubleshoot emitters that are not at least partially visible.

Once the emitters are installed on each subsystem, the final step before checking out its operation is to flush the system. The purpose of flushing is to get any bits of plastic or other debris that might clog an emitter to pass out at the end of the line. You should flush the system one line at a time—a line is defined here as a single run of PE that terminates in an end cap. Remove the end cap, then run water through the system, preferably at fairly high pressure, for at least a minute. Remove the end cap on the next line, replace the end cap on the first line, and repeat the procedure until all the lines have been flushed. Be sure to turn off the water before replacing the last end cap. After finishing the flushing, you need to run each subsystem and systematically check all the connectors and emitters for leaks. Tie a brightly colored piece of string at any problem spot, then turn off the system, repair the problems, and recheck. Once each subsystem is leak-tight, you can use your drip system to water your new plants.

# Preparing the Planting Areas

If you created any raised beds or terraced areas as part of your site modification, these beds will already be roughly in place. The remainder of your planting area will be more or less weed-free, thanks to your pre-planting weed control efforts, but otherwise pretty much unchanged. If it is a former turf area, it will present either a rich topsoil of decomposed bluegrass or a stripped down version of topsoil with no decomposing sods, depending on how the turf was removed.

It might seem like a good idea to till the planting area, but in general this is not necessary, unless you are adding soil amendments or changing the configuration of the planting bed. Tilling the surface soil will have the effect of exposing yet another crop of weed seeds on the surface, where they have the light they need to germinate, and tilled soil loses water by evaporation much faster than intact soil. But if you are adding organic amendments to your coarse-textured raised beds, these will need to be tilled in.

The question of how to find weed-free organic amendments is not a trivial one. If you make your own compost, you will know what went into the bin and what is likely to come out. Purchased compost, such as that from a municipal composting facility, is much more of an unknown quantity. The exception is composted biosolids, which is virtually always weed-free. This compost is made from municipal sewage, which might sound unattractive, but when you see this black gold, you will definitely covet it for your vegetable garden. And it does not take very much of this organic humus material to turn sand into a moisture-retentive and fertile sandy loam. The main issue with composted biosolids is possible heavy metal contamination from industrial waste, but most cities that sell composted biosolids know what waste streams enter their facility, and by law, they must frequently check their compost to make sure it does not contain anything detrimental to either plants or humans. These test results can usually be had for the asking, if you want reassurance that composted biosolids are really going to be good for your landscape.

If you do not need very much compost, bagged composted materials sold at home improvement and garden centers are usually certified as weed-free. These include composted steer or turkey manure, bark ("soil pep"), and other materials. These composts tend to be high in nitrogen—a little goes a long way. Remember, you will only be adding compost if you have created raised beds to solve a serious drainage problem and want to amend the sand to make a soil suitable for a high- or very-high-water-zone planting. Plants in low and minimal water zones will be able to handle the sand and gravel without the need for added organic matter.

Beds that have been tilled will need to be raked to smooth the rumpled surface left by tilling. The goal is not a perfectly smooth surface, which will tend to shed water if there is any slope at all. Surface roughness, on the scale of perhaps an inch, will allow water to soak in better. Raking is carried out mainly for aesthetic reasons. If the soil is very soft, it may need to be raked again to eliminate footprints after the planting is complete. Once a bed has been through the winter, it will usually be firm enough to walk on without making footprints.

## Obtaining the Plants

You have in hand a list of the plants you will need for each planting area, along with the numbers of plants needed for each species. Obtaining plants for your native landscape will not be as simple as going to a garden center to pick out petunias on planting day. Native plants are not as readily available as the annual bedding plants used throughout the country. We will try to provide you with as much information as possible to help you find the plants on your planting list, but you will want to start checking out local sources of supply well before you actually need the plants. The first place to look for native plants is any locally-owned, independent garden center in your area. These businesses have a high interest in providing quality plants to fill special needs that cannot be met through the national mass-marketing approach. Their survival depends on niche markets like the native plant market. If your favorite garden center does not carry native plants, encourage them to look for a local grower who can supply these plants for resale. Most urban areas in the Intermountain West have a few specialty wholesale growers who are trying to produce and market native plants to independent garden centers, as well as to landscape contractors and to the reclamation industry. Increasing the supply from these growers depends on letting local garden centers know that the demand is there. As demand increases, independent garden centers will undoubtedly become the most reliable places to find native plants for sale.

If you can't find your plants at garden centers, the next step is to look for nontraditional sources. These include booths at farmers' markets, native plant society sales, and botanical garden sales. The Internet is also a great resource. State and local native plant society web pages may list local suppliers, as does the website of the Intermountain Native Plant Growers Association. Sites that focus on water conservation, such as the websites of many water conservancy districts, may also have sections on native plant suppliers. Many of these suppliers are very small growers who have little visibility other than through these websites. Mail-order nurseries are another possibility. The resource section includes a list of websites to visit.

As a last resort, you could consider growing your own plants from seed. If you have a green thumb, growing natives, especially perennials, is not too much harder than growing vegetable transplants. There are many sources of native seeds listed on the Internet, and you may be able to find a local workshop, such as one run by a native plant society, that would give you the basics on propagating natives from seed as well as providing the seeds.

Keep in mind that native plants, particularly shrubs and trees, may take longer to produce in containers in the nursery than conventional landscape plants, and they will accordingly be somewhat more expensive. Bigger is not always better when it comes to container planting stock, however. Especially for perennials, small planting stock, for example, in tubes, is just as good as stock in four-inch pots or gallon containers. Do not think that you need to buy a big plant in flower in order to get a quality plant. Small plants often root in much faster than bigger plants, and most native perennials grow quickly once in the ground. A year later, you will not be able to tell the difference between a plant that came from a tube and one that came from a gallon container. It will generally be easier to find native plants in smaller container sizes, as the time investment to produce these plants is much less. This should also help to keep the cost down, as plants in larger containers are almost always more expensive. For shrubs, a sturdy plant in a gallon container is usually just as good as one in a larger container. Even trees often grow faster if planted in the ground at a smaller size. Some people want to buy big plants to get that "ready-made" landscape look soon after planting. But it is usually better to buy smaller plants and to be a little bit patient. Five years later, a landscape planted from small tree and shrub stock will usually look better and more "filled-in" than one planted from large stock, which often takes a lot longer to get started growing in the ground.

It may happen that one of the plants on your planting list is just not available, no matter how hard you look. If that is the case, you will need to find a substitute by revisiting the Plant Palette or looking for other alternatives. Or you could just leave the planting spots vacant for now and try again next year. It is not necessary to get all the plants into place at the same time, as long as new transplants into an established planting are hand-watered as needed to get them rooted in.

Arrange to purchase your plants well in advance if you can, but try to bring them home only a few days before planting. Inspect all pots for weeds. A simple pot-weeding session can save you a weed control headache later. Once you have the plants onsite, they need to be held in the shade and kept well watered. By holding your plants in shade outdoors (and even our most shade-intolerant natives can take a few days of shade) and watering them frequently, you will toughen them up somewhat for transplanting but will not expose them to undue stress from rapid drying. Native plants are generally very tough. They usually transplant easily, without showing any obvious signs of stress, especially if the planting takes place in reasonably cool weather.

## Laying Out the Planting Plan on the Ground

We like to lay out the planting plan using markers such as surveyor's flags or pieces of plastic or stiff paper labeled with plant names and tacked to the ground with nails, before actually carrying the plants to their assigned destinations. This part of the process is very satisfying, because you are taking your ideas from paper and bringing them to life. Marking the planting locations makes it easier to make any last minute spacing adjustments visually, with a minimum of pot lugging, before any irreversible commitments are made. Labeling the planting locations helps especially if the project is large and there are lots of people planting. Laying out the planting is a job for the person in charge, whereas bringing the labeled plants to their respective labeled locations and planting them can be work for any of the other people. If the project is large, it is better not to bring all the plants out of the shade at once, but to lay the plants out and get them planted into the ground one section at a time.

## Planting

Planting holes for plants in tubes or pots.

The first step in planting is to make the planting hole. The shape and size of this hole will, of course, depend on the shape and size of the pot that contains your plant. The depth of the planting hole needs to be such that the top of the root ball sits at the correct depth (one half to one inch below ground level). At this depth, it can be covered with soil without creating a mound, which would tend to make the root ball dry out easily. The width of the planting hole will depend on soil drainage. If the soil is well drained, just make the hole wide enough to contain the root ball. However, if the there is any problem with drainage, double the hole width to create a zone of loosened soil that allows water movement away from the plant roots. Once you have a planting hole, it is a very good idea to fill the hole with water, then allow most of the water to drain, before placing the plant in the hole. This is a much more effective way of providing water to the newly-developing roots than trying to top-water after planting.

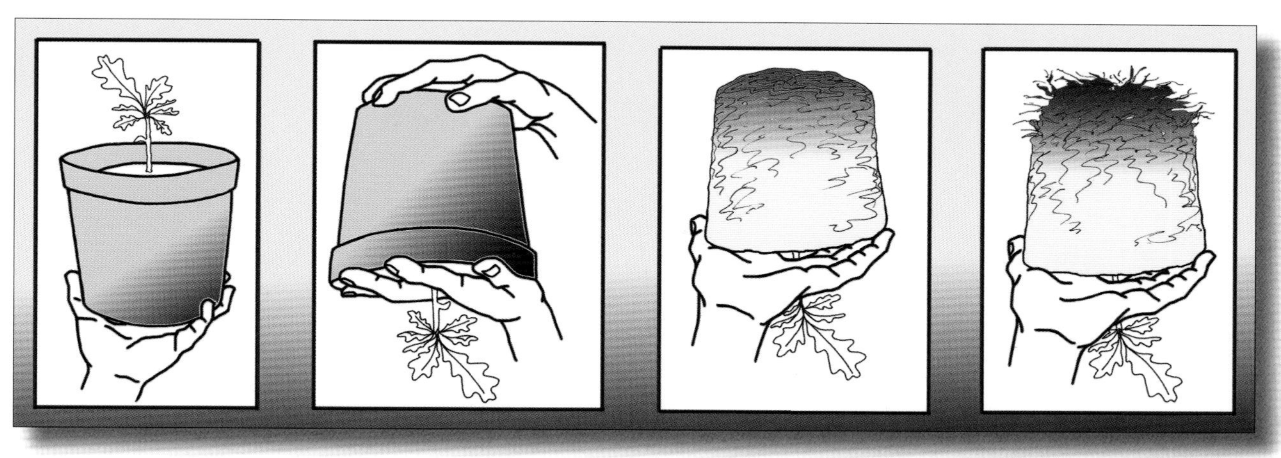

Getting the plant out of the pot.

Make sure that the plants are well watered prior to planting. To remove a plant from a tube container, turn the container upside down, knock the top edge of the container against something hard (like the edge of your shovel) to loosen the plant, and then shake it out ketchup-style, supporting the soil surface with your hand. Do not pull the plant out by the stem. If it is not fully established, it may be ripped out and lose most of its roots. The same strategy applies to plants in larger pots, except that you can usually wiggle the pot off without the knocking and shaking steps. For root systems that have been overgrown in the container, there will be roots circling around the outside of the root ball. We recommend that you loosen up the outside of the root ball and pull or cut some roots free. For plants in tubes, you can cut or pull off the tip of the root ball. You can be quite rough with the root system at this point without damaging the plant. Planting a smooth root ball into soil is much worse for the plant, because it creates gaps that are difficult for roots to cross. A rougher surface creates more surface area and more loosened roots in contact with the surrounding soil, speeding establishment.

Slip the plant with its loosened roots into its hole, being careful not to bend any major roots at right angles. This is called J-rooting, and is especially harmful to trees and shrubs. Fill in with soil from the hole, making sure to tamp it down to remove any air pockets around the root ball. Use the remaining soil to create a ring-shaped well, or moat, around the planted area, at a distance less than twice the diameter of the root ball. This moat will help hold water in place, especially water that is applied by hand during establishment, until it can penetrate the root zone. This is especially important for trees and shrubs on slopes. The final step in planting is to "water in" the plants. If you filled the planting holes with water prior to planting, the plants will not be suffering, and you can wait until you have the whole area planted before watering in the plants. The main reason for watering at this point is to settle the soil around the plant to

In this natural wash, notice how much the rocks vary in size, from boulders to grains of sand.

This example of rock mulch uses three sizes of rock.

eliminate any remaining air pockets. As the soil is already saturated, it will not require much water to get the plants settled in. Now is the time to install the emitters on your random-spacing drip system, using the guidelines described under drip installation. You can then use the drip system to water in your plants.

## Mulching

Once you have planted and finished installing your drip system, mulching is the final touch to complete the landscape. Mulching serves the dual purpose of looking good and discouraging weed growth. To suppress weeds, and to make sure the soil is covered, use at least three but no more than five inches of mulch. Mulch that is too thin tends to act as a seed bed for weeds. Applying an inch of mulch to a thousand-square-foot area requires about three cubic yards of mulch material. If your area is a hundred square feet, and you want to apply three inches of mulch, you will need about a cubic yard of material (100/1000 × 3 cubic yards × 3 inches = 0.9 cubic yards). You do not need to mulch immediately after planting. Avoid being in a rush when applying mulch, especially around small plants. Be careful to leave a sizeable mulch-free area around each plant. It is very easy to bury and smother small plants with mulch. In fact, it is often best in perennial plantings to apply only an inch at the beginning, and give the plants some time to get a little taller before topping up the mulch to three inches. At the very least, leave the area inside the "moat" you made after planting mulch-free. It is usually not a good idea to use landscape fabric (weed mat) under mulched planting areas. Few natives other than water-loving plants thrive with "wet feet," and weed mat under mulch tends to be a little too good at conserving water near the surface. Weed mat is also hard to manage in perennial beds because of the need to replace plants over time. It definitely gets in the way when replanting. If you do use landscape fabric, be sure to put drip lines on top of it, not underneath.

As part of the design process, you targeted different planting areas for either organic mulch (e.g., bark) or an inorganic rock mulch. Organic mulch can be spread uniformly. To make rock mulch more appealing, increase its complexity and mystery by using gravel and cobbles of different sizes and colors in a naturalistic, drifting pattern and in combination with larger rocks.

## Watering to Establish

All plants, even the most drought-hardy desert plants, need regular supplemental water during establishment. This is because the deep, extensive root systems that enable these plants to survive on little water need time to develop. A container is an artificially favorable environment that permits a plant to make more top growth than its pot-limited roots would be able to support in nature. When you first put the plant in the ground, it still has this lopsided configuration of too much top for the roots to support. In time, the roots will grow out into the soil, the process called "rooting in." But at first, the plant will need to be watered almost as often as if it were still in the pot.

How long the process of rooting in will take depends on several factors. Perennials usually root in faster than shrubs, and shrubs root in faster than trees. In general, the larger the plant is at the time of planting, the longer it will take to root in. If you plant in the fall, perennials and small shrubs will be well rooted in by the end of the following spring. Many large shrubs and trees will require supplemental water at least through the first full growing season. If you plant in the spring, perennials and small shrubs should be rooted in by the end of the first growing season, whereas many large shrubs and trees will require an additional full growing season of extra supplemental water. Trees may not show signs of water stress if inadequately watered during establishment, but their growth rates will be much reduced. If you are interested in a rapid growth rate, then extra water during the first two growing seasons is a good idea for trees.

How often and how much to water during establishment also depends on both the plants and the circumstances. If you plant when the weather is cool (spring or fall), fill the planting holes with water, and water in the plants, then you only need to water about once a week for the first month, and even less if there is substantial natural precipitation. You can cut this back to once every two weeks for the second month, then start the regular watering regime for plants that will receive regular irrigation, i.e., the plants in water zones that will receive supplemental water. For plants that will not receive regular irrigation, water at least once a month for the remainder of the growing season. Watering beyond the second month, or sometimes even the first month, will generally not be necessary for fall plantings. For large shrubs and trees in unirrigated zones, water once a month during the second growing season, starting when the weather warms up and the soil begins to dry. If you plant when the weather is hot, a practice that is not recommended but that works better for natives than for many traditional garden plants, you will need to water more often, especially during the first month. Watering as often as every second day may be necessary. In general, when watering to establish plants, be sure to add enough water each time to soak below the root ball, so that the roots are encouraged to penetrate deeper into the soil.

*Indian paintbrush*

# How to Care for Native Landscapes

Chapter 5

One of the main motives for using native plant landscaping in place of traditional landscaping is the idea that the native landscape will require fewer resources and less maintenance but will still look as beautiful as a traditional landscape. Beauty is in the eye of the beholder, but we certainly believe that a well-designed and well-maintained native landscape is far more beautiful than a vast expanse of lawn punctuated only by a row of junipers and a concrete-rimmed bed of petunias. And it is easy to demonstrate that a native landscape will thrive with much lower resource inputs—less water, less fertilizer, fewer pesticides. But the question of maintenance requires closer examination.

Native landscapes do require less maintenance than traditional landscapes, but more importantly, the maintenance they require is strategic maintenance. Native landscape maintenance is not like the weekly grind of watering, spraying, fertilizing, and then mowing to remove the excess herbage generated by all that watering and spraying and fertilizing. It is not even like the regular and frequent attention that you need to give to a well-watered, well-fertilized vegetable garden. Native landscape maintenance is strongly seasonal and often quite flexible, much more flexible than traditional landscape maintenance. Sometimes weeks will go by with little to do in your landscape but enjoy it. But when the time comes to water, or to weed, or to prune and deadhead, native landscapes, like all landscapes, benefit from some concerted attention. Low maintenance is not "no maintenance," just as xeriscape is not "zero-scape"—as it is often misstated in real estate ads in drier parts of the country. Basically, the maintenance tasks are watering, weeding, managing plant appearance, maintaining the hardscape and the irrigation system, and managing the mulch. This list is much like the list for any ornamental garden, but in a native landscape, the magnitude of these tasks is usually much reduced and focused over shorter time frames than in traditional landscaping.

## Hardscape and Watering System Maintenance

Maintaining hardscapes in a native landscape is essentially the same as maintaining hardscapes in any landscape. Depending on the materials used, regular staining or painting may be necessary, sand between the bricks or stones of a path may need to be replenished, and fences, walls, or other structures may need to be repaired. Each spring, you will need to inspect your hardscape and determine what maintenance and repair activities are necessary. Maintaining a drip irrigation system also requires regular attention. At the end of the season, you need to remove the end caps from the drip lines and let them drain, in order to minimize the chance of freezing damage. Before operating the system in spring, you will need to re-install the end caps, flush each line, inspect all the connectors and emitters for leaks, and repair any problems you encounter, just as you did when first installing the system. Usually any problems encountered are minimal, as drip irrigation systems can operate for many years with little or no replacement of parts.

## Long-term Watering

Once your plants are established, it is time to begin the long-term watering regime. For water zones that are the same as the zone at your site or drier, this basically means little or no watering at all unless the year is exceptionally dry. For zones that require supplemental irrigation, the water will hopefully be added in a relatively painless manner, through an established drip irrigation system. Watering with drip irrigation is very easy once the system is in place.

In water zones without a permanent irrigation system, it requires some judgment to know when supplemental water is necessary. People in the Intermountain West are generally very aware of the status of the "water year" in their area, which starts on October 1 each year. The nightly TV weather report will often include a statement something like, "The mountain snow pack for the water year is only at 60 percent of normal." This is cause for worry, because it is the melting snow pack that fills the reservoirs that in turn provide the culinary and landscape water for most residents of the area. What "60 percent of normal" means is that the amount of water that has fallen as snow or rain from October 1 up to that date is 60 percent of the average amount that falls by that date. Totals that are far below average during the spring months in the mountains usually correspond to exceptionally dry conditions in the foothills and valleys as well. This can be a cue that your unirrigated landscape will need some supplemental water. Checking the precipitation amounts for the water year for a nearby weather station at a comparable elevation can tell you how far below average local precipitation has been. If the winter and spring have been dry, with precipitation less than 80 percent of average, it is a good idea to apply two or three inches of water at the beginning of the summer, say in late May. This watering can be carried out at night with a portable sprinkler over a two or three day period, or each plant can be hand watered two or three times with the equivalent of an inch of water over the area of its root system. This will substitute for the recharging of deep soil water that usually occurs with adequate winter precipitation, and will usually be enough to get your plants through the summer in good shape. If really grim heat and drought conditions persist, you can water once in midsummer, again applying at least two inches.

To determine how much and how often to irrigate in the water zones in your landscape that require regular irrigation, you calculate the total number of inches you need to add, then divide this number by two, because you want to add about two inches each time you water. This is to be sure that the water penetrates deeply into the soil. The total number of inches divided by two equals the number of times you will need to water during the course of the growing season. For example, if you are in the semi-desert with eleven inches of annual precipitation and you have a mountain planting in a high water zone, you need to add fourteen inches of water to top the water up to the twenty-five inches that mountain zone plants expect. If you add two inches each time you water, you would need to water seven times over the course of the growing season, starting when the soil starts to dry out in early summer. It is usually not necessary to begin supplemental irrigation until around the first of June in the semi-desert zone, and by mid-September, the weather has cooled off and autumn storms have often arrived. Seven irrigations spread across this fourteen-week period means that you would water approximately once every two weeks. For a foothill planting in this same semi-desert zone, you need to add about nine inches, which would be about five 2-inch irrigations. This means you would water about once every three weeks in your medium water zone. Rounding up the amounts so that your watering schedule is on the basis of weeks makes it easier to keep track of when you need to water.

If the winter and spring have been exceptionally dry, the best way to help out the plants in your higher water zones is to add extra water at the beginning of the

## Watering Guidelines for Established Plantings in Different Landscape Watering Zones

Watering guidelines are based on average annual precipitation in inches at the site.

| Site Precipitation Zone | Average Annual Precipitation (inches) | Inches of Water to Add in an Average or Better Water Year for Plantings in Each Landscape Water Zone | | | |
|---|---|---|---|---|---|
| | | Minimal | Low | Medium | High |
| Desert | 6 | 4 | 10 | 14 | 20 |
| Desert | 7 | 4 | 8 | 14 | 18 |
| Desert | 8 | 2 | 8 | 12 | 18 |
| Desert | 9 | 2 | 6 | 12 | 16 |
| Desert | 10 | -- | 6 | 10 | 16 |
| Semi-desert | 11 | -- | 4 | 10 | 14 |
| Semi-desert | 12 | -- | 4 | 8 | 14 |
| Semi-desert | 13 | -- | 2 | 8 | 12 |
| Semi-desert | 14 | -- | 2 | 6 | 12 |
| Semi-desert | 15 | -- | -- | 6 | 10 |
| Foothill | 16 | -- | -- | 4 | 10 |
| Foothill | 17 | -- | -- | 4 | 8 |
| Foothill | 18 | -- | -- | 2 | 8 |
| Foothill | 19 | -- | -- | 2 | 6 |
| Foothill | 20 | -- | -- | -- | 6 |
| Mountain | 21 | -- | -- | -- | 4 |
| Mountain | 22 | -- | -- | -- | 4 |
| Mountain | 23 | -- | -- | -- | 2 |
| Mountain | 24 | -- | -- | -- | 2 |
| Mountain | 25 | -- | -- | -- | -- |

1. The sum of natural precipitation and added water for each landscape water zone equals the high average value for that zone (10 inches for desert, 15 inches for semi-desert, 20 inches for foothill, and 25 inches for mountain).
2. Two inches are recommended at each irrigation, so the total irrigation amounts are adjusted upward to represent multiples of two.
3. To calculate the number of times to water, divide the total number of inches by two.
4. If precipitation at the site exceeds the water requirement for plantings in a particular landscape water zone, the irrigation value is left blank.

season, just as you did for the plants in your unirrigated zones, to help recharge the deep soil water. You could start watering a couple of weeks earlier, and add two or three inches of extra water before beginning the regular irrigation season. If the summer is brutally hot and dry, you could add an extra irrigation or two some time during the hottest part.

It is also perfectly acceptable to let the plants tell you when they need extra water. This will not be as obvious as the wilting of a tomato plant, but with experience you will learn to recognize the signs. Resist the urge to add water in amounts greatly in excess of those recommended, however. Remember, there is such a thing as too much water. For example, penstemon plants that are dying of vascular wilt diseases aggravated by overwatering look a lot like penstemon plants that are wilting from lack of water. And be aware that yellowing leaves usually mean that your watering is excessive, not that the plants need more water.

If the autumn weather is much drier than usual, you may need to add an extra irrigation at the end of the growing season. Plants usually overwinter better and are less susceptible to frost damage if they go into the winter in a well-watered condition.

## Staying Ahead of the Weeds

If you took care of serious weed problems before planting, installed an irrigation system that does not water the interspaces between the plants, applied a thick layer of mulch over most planting areas, and used path materials that keep weeds out, your weeding time should be kept to a minimum. Ongoing weeding is like locking your car door. It doesn't take much work, but you have to be vigilant and remember to do it. Otherwise, undesirables may enter and get the upper hand when your back is turned.

Most of the weeds you see in your planting will be from seeds blown in from the yards of your less zealous neighbors. You may not be able to do much about weed seeds blowing in from adjacent properties, but you can prevent most weed seed emergence by mulching heavily and targeting the application of water. Still, some weeds will always appear, especially near the plants, where the mulch is often thinner and the water supply steadier. The first and best option is always manual cultivation—otherwise known as pulling weeds. If all your other weed deterrents are in place, this should not be too hard. The motto is to weed early and often, and never to let any weeds go to seed in your planting. To allow the weeds to go to seed multiplies your problems exponentially for the next year. View weeding as a chance to visit your plants and spend some quality time with them. You will be surprised at how pleasant this can be, as long as the weed problem is never allowed to escalate to the point where it discourages hand weeding as a control method. If the worst happens—for example, while the twelve-year-old neighbor is tending your yard while you are off on a summer cruise—you may have to resort to more draconian methods. These methods may also be necessary if established perennial weeds appear in your planting.

Herbicides are the last resort for weed control, but they may sometimes be the most practical solution. Post-emergent herbicides are sprayed onto plants that are already growing. We recommend them for use as a spot spray to control weed infestations that have appeared in spite of preventative measures. For urban landscapes, there are few post-emergent herbicides that are available and cost-effective. The most universal post-emergent herbicide is glyphosate, the product we recommended for removing lawn. It is nonselective, and works well on many annual and perennial weeds because it is a systemic chemical that moves from the leaves to the roots and kills the whole plant.

Another common and readily available post-emergent herbicide is 2,4-D. In contrast to glyphosate, it is selective, controlling only broadleaf (non-grass) weeds. For this reason, it is best suited to broadleaf weed control in lawns. As mentioned earlier, a mixture of equal parts glyphosate and 2,4-D is one of the few combinations that may actually take out field bindweed. If you have a new or persistent bindweed infestation in your planting, this is one approach that has a chance of working. There are some novel herbicides coming down the pike that may be more effective for field bindweed control, but these are not yet available to homeowners.

Spot spraying with post-emergent herbicides is quite straightforward. Follow the label instructions to the letter, including the instructions for disposing of any leftover herbicide and empty herbicide containers. One problem with spraying herbicides in established native plantings is that native plants in general are very sensitive to these chemicals, and it is easy to damage or kill your plants in the name of helping them escape from weeds. Always pick a calm day and don't try to spray weeds taller than two feet, because the chance of drift is greatly increased when the herbicide is applied farther above the ground. Another option for applying herbicides to individual weeds is to manually spread the herbicide directly onto the leaves. Wearing a disposable rubber glove, dip your hand into a container of herbicide and rub the herbicide onto the leaf surfaces. This method works even better if the herbicide is mixed with a little dish detergent, which helps it stick onto the leaves. This eliminates drift and can work when the weed is in close proximity to a native you are trying to rescue. Another option is a wick applicator, available at hardware stores.

Pre-emergent herbicides are designed to prevent the emergence of germinated seeds. These are often fairly expensive herbicides, and most are effective on only a limited range of weed species. One commonly available pre-emergent herbicide is oryzalin. We recommend using pre-emergent herbicides only as a temporary measure to control potentially severe weed infestations resulting from the dispersal of large numbers of weed seeds into the soil. This can happen if a dense stand of weeds disperses its seed crop before the plants can be removed. We strongly advise against constant use of pre-emergent herbicides. Many weeds can evolve resistance to specific pre-emergent herbicides over time. Like antibiotics, these chemicals should be used as sparingly as possible, so that they maintain their usefulness over the long term.

Applying pre-emergent herbicides requires somewhat more attention to detail than spot spraying with post-emergent herbicides, because continuous coverage of the soil surface is necessary for the chemical to be fully effective. To achieve even coverage, it is best to use a backpack sprayer with a flat fan spray, though smaller hand-held sprayers can also be used. Pay special attention to bare soil areas where weed germination and emergence are most likely. In theory, the herbicide does not damage established plants when used according to instructions. But it is probably best to avoid dousing the leaves of your natives if you can, as few herbicides have been tested on any plants other than crops and traditional ornamentals. Most pre-emergent herbicides need to be watered into the soil with at least an inch of water in order to be activated. In an unirrigated or drip-irrigated native landscape, this would require the use of portable sprinklers or a herbicide application just before a rain. Most remain active in the soil for three or four months.

An organic alternative to chemical pre-emergent herbicides is corn gluten meal. This byproduct of feed corn processing can kill germinated weed seedlings before they emerge, just as a chemical pre-emergent does. It has no toxicity to established plants or other living creatures, and it biodegrades fairly quickly, remaining active in the soil for about four to six weeks after application. It too requires watering in to be effective, but it is applied in a granular formulation that can wait on the ground for the same rain that triggers the weed seeds to germinate. Corn gluten meal can be applied with a fertilizer spreader. It is effective against a wide range of weeds, and the weeds do not evolve resistance to it. Desirable plants are safe once they develop two or three true leaves—the herbicide only affects newly germinated seedlings. One possible drawback to corn gluten meal in native plantings is that it acts as a nitrogen fertilizer. It is sold under several trade names both in garden centers and over the Internet.

## Maintaining Mulch

As your first line of defense against weeds, mulch is an important element in your landscape. Unfortunately, mulch does not last forever, so a regular program of mulch replacement or replenishment is a necessary component of long-term maintenance. If you use bark mulch, it will need to be replaced approximately every three years, because over time weeds can root into the decomposing mulch. Rock mulch will not need replacing *per se*, but a danger with rock mulch is the buildup of organic matter that can ultimately form a seed bed for weeds. In effect, the gravel becomes incorporated into the surface of the underlying soil. Raking over the gravel mulch can help to knock the organic material to the bottom of the rock layer, increasing the effectiveness of the mulch. This is fairly labor-intensive but will only occasionally be necessary. If water is plentiful, an alternative tactic is to wash the organic material to the bottom of the gravel layer. Another solution is to add more gravel on top, if there is no problem with raising the level of the ground surface a few more inches. Organic matter generally only accumulates in the gravel in areas where there is a significant source of debris from leaf fall, and these planting areas are more likely to have organic mulches than gravel.

## Keeping Natives Looking Their Best

Another appeal of native landscaping, in contrast to traditional landscaping, is the idea that a planting is a permanent feature. Replacing annual bedding plants every spring is a way of life for traditional gardeners, and bedding plants are the lifeblood of the mainstream nursery industry. People who want to get off this treadmill look to perennial plants that can provide value for many years, and most natives are perennials *par excellence*. But native perennials vary in their life expectancy, and the life span of a native plant is often dependent on growing conditions. No plant lives forever, except possibly clonal plants that reproduce by runners or rootstocks, and these are usually not plants that are well behaved in the landscape.

Native plants often do not live as long in human-created landscapes as they do in nature. There are many reasons for this. Sometimes the good life is so good that the plant invests too much into flowering and reproduction each year, and spends itself in a few years. It achieves high seed production this way, but that is not your goal—you want the plant to live a long time. The too-good life can also make the plant less resistant to various stresses. For example, if the fertility is too high, a plant may become oversized and lush, but it also becomes less able to tolerate drought or cold. And, as we have mentioned before, the too-good life of too much water and organic matter can often make normally tough and disease-resistant plants susceptible to fatal soil-borne diseases.

You can do your best to provide the conditions necessary for a particular plant to live a long and healthy life. But sometimes an unforeseen problem will rear its head. Most native plant gardeners like to experiment, and are not daunted by a little failure. So the bottom line is that some plants in a native planting will need to be replaced, particularly perennials. Most native shrubs and trees are long-lived once they have survived a year or two—problems with these plants usually surface right away. Removing and replacing any plants that succumb is an important maintenance activity. If the plant seems to have met an unnatural end, it is best not to place another plant of the same species, or often even of another species, directly into the same spot. Soil-borne diseases multiply in the disintegrating roots of sick plants, and another plant in that spot may have a high probability of contracting the same disease.

Another maintenance activity is deadheading, or trimming back the flowering stalks of perennials once they have finished flowering and before their seeds are ripe. This is mainly for aesthetic purposes, but it may sometimes cause another round of flowering in a species that normally flowers only once, an added bonus. This kind of trimming can be carried out during the course of the growing season. Another round of stalk trimming and general clean up is usually necessary in late fall or early spring, when the plants initiate new growth.

One good reason to deadhead flowering stalks is to prevent self-seeding. Unlike artificial hybrids and tropical species grown in short-season climates, native plants are generally very good at setting seed. After all, the purpose of a flower is to make it possible for the plant to reproduce itself through seeds. Long after you have solved all the weed problems in your planting, the native species you have planted will be continuing to volunteer from seed. Dealing with volunteers is a lot like weeding, only harder, because you like these plants. The best way to avoid this conflict is to remove seed heads before they are ripe. The alternative, to give up control and just let the good times roll, is discussed in the next section.

Woody native plants also require periodic attention. At a minimum, you will need to remove dead wood, both for aesthetic reasons and for the health of the plant. Sometimes you will need to prune back shrubs that have overgrown their spaces, or trees that have developed poking branches right at eye level. You may also want to limb up trees, or remove basal sprouts, to encourage a more tree-like form in a species like mountain ash. In general, though, most native woody plants have naturally pleasing shapes that require little pruning. Shaping a plant through pruning becomes a matter of personal taste rather than necessity. Most natives are tolerant of pruning because of their long history of interaction with fire or browsing animals like deer. Few would argue that hedging by deer results in an improvement in a plant's appearance. Fortunately, people can do a more judicious job of pruning. Thinning many-stemmed shrubs by taking out stems at the base, much the way a forsythia is pruned, can often open up a shrub and make it more attractive. This is especially true for shrubs from the foothills and mountains, such as golden currant.

Some fast-growing shrubs can benefit from more drastic pruning. Many native shrubs are fire-tolerant and can sprout back from the base after complete destruction of the aboveground portion of the plant. For example, rubber rabbitbrush often grows in very coarse soils in nature, and the too-good life in the garden can make it rank and overgrown over time. Shearing the entire plant at ground level each year in the late fall or early spring results in profuse sprouting and a short, compact growth form that is very pretty. An advantage to shearing in the fall is removal of the seed heads, as this plant is a prolific self-seeder. Other shrubs that can be pruned very heavily or sheared include oakleaf sumac and Apache plume.

## Change in the Native Landscape

Earlier in this book, we emphasized the idea that native landscapes are dynamic, changing over time, both with the seasons and over the longer term. When your native landscape is new, you relish the changes as it matures. You anticipate and welcome the first flowering of the perennials, the filling out of the shrubs, the growing of the trees. Each spring you look forward to the orderly seasonal progression of budding, leafing out, and flowering. At a certain point some time in the future, you will look out and realize that your landscape now resembles the picture you had for it in your mind, and you will feel great satisfaction. At this point there is often a desire to preserve this picture-perfect state, to make the landscape static in time. This is virtually impossible. The perennials will attempt to seed themselves all over the property, and then eventually they will succumb to the ravages of time, whether or not their progeny have ruthlessly been weeded out. The shrubs may take on a ragged, overgrown look, and the trees will get bigger than you ever believed they would, even though you carefully planned for their mature size in your original design. They now shade large areas that were supposed to stay sunny forever. How do you cope with these changes? If you know from the beginning that you will need to continue to interact with the landscape and guide its development, you will have an easier time dealing with change. Some people believe that a native landscape inevitably matures, peaks,

1999

2000

2002

2005

Change in the desert garden section of the Utah Heritage Garden at Wasatch Elementary School in Provo, Utah. The garden was planted in spring 1998 and is still thriving and changing.

and declines, but your role is to provide the means for cycles of rejuvenation. If you do this, your landscape will never decline. Yes, it will change, it will not be the way it was that glorious day when you realized it had achieved picture-perfection. But it will continue to satisfy and to teach, as long as you are willing to devote your attention to the process.

Rejuvenation can mean providing opportunities for your perennials to self-sow a new generation before they die. You can direct this process by saving seed and planting it in the fall in areas cleared of mulch to prepare a seed bed. Rejuvenation can mean hard pruning to encourage directed new growth, or even shearing shrubs to the ground so that they can rise again, renewed. Rejuvenation certainly means replanting after the loss of existing plants. Take any changes in the microenvironment into account when planning to replant—you may need shade-loving plants where you once had sun-lovers, for example. Rejuvenation also means accepting that certain plants in certain places were a bad idea, and biting the bullet to move or remove them. If some supposedly well-behaved native has turned weedy on your site, just eliminate it. If a plant billed as shade-tolerant looks leggy and dejected in too-dense shade, move it to a sunnier spot and try something else in the shady spot. If penstemons die after a few years in your native soil, give up and plant something less finicky there—and try moving the penstemons to a raised sand bed. In fact, part of the fun of native landscaping is the freedom to keep making changes. Experimenting with new species is always stimulating, and you will never run out of new natives to try. Certainly, embracing change in a native landscape is far better than living with the utter monotony of lawn. If you keep an open mind about change, you will continue to enjoy your native landscape for a lifetime.

*Utah ladyfinger milkvetch*

# Native Landscape Pioneers Tell Their Stories

Chapter 6

## Return of the Natives

Phil and Judy Allen, Orem, Utah—*Semi-desert Zone*

Personally, my choice to "go native" was as much about reconnecting with favorite childhood plants and memories as it was about art or philosophical ponderings related to environmental stewardship. Still, ripping out what remained of our front lawn in 2004 felt strangely awkward. (Every other front yard on our street had copious quantities of Kentucky bluegrass, and my PhD in horticulture focused on high-maintenance turfgrasses). That said, the journey from a solid carpet of lawn to the creation of a Wasatch Front canyon landscape has been worth it in every way.

We purchased our brick rambler in 1992. At that time, the completely flat landscape was dominated by lawn that reached from the house to all property lines. The trees included exactly one of each of the following, growing as "lollipops" in the grass: Norway maple, sycamore maple, quaking aspen, flowering plum, cherry, apple, and Douglas fir. The only shrubs in the yard included four dwarf Alberta spruces (located in the front yard at the corners of the house and sides of the porch), a single lilac, and a row of pfitzer junipers along the fence in the back yard. We immediately removed the flowering plum, which was located in the center of the back lawn and conflicted with soccer.

It took just over one year to complete the initial design for our yard. The idea was to use a combination of favorite Utah and Minnesota plants—Utah where

we were raised and Minnesota where I had attended graduate school. The design included large beds for trees, shrubs, and perennials. This, we felt, would allow plenty of room for the plants to grow without seriously competing with the lawn. In the front yard, planting beds were located both near the house and as a curvy island in the center of the yard. The entire perimeter of the back yard was converted to curvilinear beds, while the center of the yard was kept open for recreational lawn use.

I then designed and installed a sprinkler system to keep the grass happy. It was actually quite impressive to watch my well-designed system at work, efficiently watering the lawn while virtually eliminating runoff onto the sidewalk and driveway. And the new back-yard sprinklers no longer watered the cedar fencing. Over time, the arcs of hard water deposits gradually faded from the fence slats.

Functional design considerations for our yard called for a number of deciduous trees to shade the house during the summer, while allowing sunshine to warm the bricks and siding during the winter months. We accomplished this by choosing relatively small trees to use near the house (bigtooth maple, Gambel oak, staghorn sumac, and water birch) and large trees for farther away (bur oak and a *temporary* Norway maple). We achieved aesthetic appeal in three ways. First, we repeated native plant clusters in multiple locations. Second, we added white fir to the clusters in order to create strong winter interest and provide balance and contrast to the deciduous trees. Shrubs and perennials, generally planted in groups of three or five, were added to fill in the gaps while the trees grew, as well as for additional variety. The ground cover we used beneath tree clusters was creeping Oregon grape, which has attractive yellow flowers in the spring and reddish leaves in the winter.

Unfortunately, the two native trees (Douglas fir and quaking aspen) had been planted directly beneath overhead utility lines, so they were cut down within a few years. But the Douglas fir made a fine Christmas tree (at least the top six feet!), as did each of the dwarf Alberta spruces. Today, only three of the original trees remain. The two fruit trees located in the corner of the back yard were kept as much for the shade to my kids' play area as for the fruit. Also, the Norway maple has been granted a temporary stay of execution because we need the shade until our new forest can replace this vital function.

As previously mentioned, the original design included several intermountain natives. Gambel oak and bigtooth maple were obtained from a friend as six-inch seedlings. Later, additional oaks were added as acorns. White fir and Douglas fir were purchased as tubelings, and western water birch and netleaf hackberry were purchased in one-gallon containers. Big sagebrush, fernbush, oakleaf sumac, curlleaf mountain mahogany, green Mormon tea, Rocky Mountain maple, basin wildrye, and western mountain ash were either purchased as seedlings or donated by friends from the Utah Native Plant Society. The Oregon grape starts were dug up on nearby Forest Service property, under a special removal permit obtained during a road-widening project.

We generally plant trees in early spring or autumn. Our soil is nearly filled with rocks, ranging from the size of your fist to that of a football. But my trees obviously love it! In contrast with numerous written and verbal claims that native intermountain trees are inherently slow growing, my experience has been that they grow as quickly as many nearby non-native plants installed at around the same time. For example, oak seedlings planted in 1995 are approaching fifteen feet in height just twelve years later. The water birch and bigtooth maple are even taller, and one limb of the hackberry shot up to twenty feet before I pruned it. My personal view is that our native trees perform best when they have a large root system, especially at an early age. I once dug up a Gambel oak seedling to discover that, at the four-leaf stage (less than three inches in height) the taproot was already more than a foot deep! Nursery production practices are just beginning to incorporate an awareness that native plants from semi-arid habitats benefit from much larger root volumes than traditional containers allow. Because all our trees were planted at a small size, the problem of trying to transplant a tree with insufficient root volume was avoided.

In the front yard, the trees and lawn tolerated each other fairly well, at least for a few years. Eventually, tree roots reached the edges of the driveway and sidewalk, turning to run parallel to the concrete. It became increasingly difficult to apply enough water to keep the grass green next to the concrete, and I began to tire of the need to overwater most of the lawn in order to keep the one foot next to the concrete happy. But what was the alternative?

On my 40th birthday, my wife and I climbed a large talus slope in nearby Rock Canyon. I noticed that where the forest and talus met, rocks dominated the soil surface beneath the trees for several feet. Additional forays into the steeply sloped, tree-covered canyon faces confirmed that the forest floor on many sloped areas is indeed covered with rocks. So the idea for an alternative to lawn was born. Additional benefits were that weeds would virtually be eliminated, and in this context, at least, the rocks were aesthetically appealing. We obtained a permit from the Bureau of Land Management to harvest two tons of rock from a talus slope located twenty-five miles from our home. The last mile to the rock pile was on a marginal dirt road that might have deterred us, were it not for the fact that we had already severed the lawn sprinkler pipes in a few locations. We eventually purchased a total of three permits and hauled eight pickup and trailer loads of rock back to the yard. To contrast with the ground cover of rocks, we purchased yellowish river stones for the dry stream bed that now meanders through the forest. And the walking path that connects our front sidewalk to the porch was built by recycling sandstone from our living room fireplace, which was removed during a remodeling project.

As with other journeys into the unknown, we have learned a great deal from our landscaping experience. If we were starting over today, we would have passed on the bur oak. Gambel oaks are just as lovely and can be coaxed to grow quite tall through close spacing—you then thin some trunks as they mature. We

planted a few clusters of ginkgo around the front yard as a substitute for quaking aspen, and in retrospect I wonder if aspen wouldn't have worked just as well. A healthy forest often includes trees of various ages, and we regularly remove trees that are growing too tall for the scale of our home. Each year we cut down a fir tree for Christmas. We spread a few new acorns and bigtooth maple seeds around the yard each year, and we plant new fir seedlings every few years as well, so there is a continual supply of new seedlings for the future forest. We could easily renew the quaking aspen by eliminating trunks as they become diseased or outgrow the yard.

After killing the lawn, we should have removed all the dead sod wherever we were going to plant native perennials. The high organic matter that was left over led to fertility levels that are too high for penstemons and gilia, and even the sagebrush appears too lush and needs pruning to keep it from lodging (flopping down).

While we purchased components for a drip irrigation system and have much of it plumbed already, we have never felt the need to actually finish and operate it. French drains originate wherever downspouts come down from the roof and lead to key tree clusters. While shoveling snow in the winter, I toss it onto areas where more water might be needed. In the summer, we water plants by dragging a hose a couple of times each year. We have learned, however, that watering the wildflowers occasionally (every few weeks when in flower) can significantly extend the blooming season. Some years we do, others we don't. However, we specifically selected plants for our yard that could survive for two or three months without being watered.

We initially worried that our unconventional landscape would offend the neighbors. But the same views we enjoy from every window of our house are enjoyed by the neighbors in reverse. And children who walk or ride their bicycles down the sidewalk feel compelled to enter the yard, to walk across the bridge that crosses the stream bed, to feel the rocks. As a boy, I loved to picnic with my family in wild places. In the canyons I climbed trees, threw pine cones, and gathered acorns and painted them with faces. In the deserts I chased rabbits through sagebrush and green Mormon tea, and even today the fragrance of these wonderful plants takes me to a place where I feel free and playful. On those days when I don't have time to actually go to a wild place, my yard works just fine.

## Sharing a Patch of Sagebrush

Jan Nachlinger, South of Reno, Nevada—*Desert Zone*

Native plant landscaping at the western edge of the Great Basin is neither very difficult nor worthy of praise when much of our yard had not been significantly altered from the natural shrubland landscape. In 1990, I was thrilled when we stumbled on the mostly intact 1.25 acres in a rural area south of Reno—the house was nearing completion, and the local independent home builder had only disturbed the building and driveway footprint. The presence of an intact native plant community was among the chief reasons that we decided to buy the house, as I knew I was well on my way to having the kind of yard I wanted.

The property is at 5,500 feet in elevation, on a northeast-sloping alluvial fan adjacent to eastside foothills of the Sierra Nevada. The local shrubland is dominated by Wyoming big sagebrush and antelope bitterbrush, with desert peach, spineless horsebrush, and woolly mules-ears. In summer, spurred lupine, Anderson milkvetch, Bruneau mariposa lily, foothill deathcamas, tall woolly buckwheat, and dwarf purple monkeyflower adorn the shrubland. Yes, there is cheatgrass. It is a persistent problem requiring attentive weeding—especially in wetter years—but, at least it seems to have lessened its stronghold in the shrub interspaces, due to competition from perennial grasses (squirreltail and needlegrass) that I broadcast-seeded from a locally collected seed bank.

Planting trees was our initial priority, because we wanted shade and a vertical dimension added to the three- to five-foot-tall shrubland. I knew that the native trees I wanted would be slow growing, so I scoured a number of local nurseries for relatively large trees and had great fun on a rented bobcat digging holes for their planting. The soil is fertile though very rocky. I designed the tree layout to coincide, for the most part, with the larger ecological context of their natural distributions. That is, I planted eastside Sierran trees (Jeffrey pine, ponderosa pine, incense cedar, western white pine, and curlleaf mountain mahogany) on the west side of the property and Great Basin trees (singleleaf pinyon, Great Basin bristlecone pine, Rocky Mountain juniper, Rocky Mountain lodgepole pine, Engelmann spruce, and Rocky Mountain ponderosa pine) on the east side. For color among the Sierran trees I planted western mountain ash, western columbine, Sierra spirea, and firechalice as associates. Planted associates for the Great Basin trees and sagebrush included oakleaf sumac, creeping Oregon grape, Nevada buckwheat, Lewis flax, showy penstemon, and firecracker penstemon. In an effort to salute the Mojave Desert of southern Nevada, I planted Apache plume, Gambel oak, fernbush, and Bridges penstemon at the southwest corner near a Rocky Mountain ponderosa pine—a

virtual Spring Mountains landscape in miniature. We installed drip irrigation lines throughout to water the trees, as the site averages only eight or nine inches of precipitation annually.

We chose to plant an aspen "grove" in lieu of lawn and located it centrally in the south part, where there was a natural opening in the shrubs. It is in our view through the front south-facing windows. The grove is modeled after eastside Sierran stands, and in addition to several quaking aspens, it includes mountain alder, two Sierran lodgepole pines, and a volunteer arroyo willow which came in just a year or two after we started drip-watering the grove. Because willow seed is short lived, I imagine it drifted in from one of the nearby streams, where it is abundant. I purposely kept the middle of the grove open so that we could have a small shady respite on hot summer days, and it is one of the yard's delights to lie or read among the aspen gently trembling in the breeze.

Most of the aspen are healthy and suckering; however, one died about five years ago, and two more have been losing vigor for several years. I may not be watering some of them adequately to reduce their summertime transpiration stress, but since I obtained them from several sources, some appear to be genetically better adapted to their artificial home in the sagebrush. As with most of the other native plants that I have added to the yard, if they don't survive with my minimal caregiving, I move on to less finicky options. Lately, the grove has had a

Cooper's hawk occasionally perched in it, which makes me feel like it's a pretty successful imitation of a natural stand.

Early on I brought in many cubic yards of decomposed granite. It took quite a while to shovel it along the several pathways winding through the property to different interest points. But the effect is a nice and easy natural path, and it doesn't take very much time to pluck out the sagebrush and bitterbrush seedlings that inevitably germinate each year. What takes more time is the annual removal of dead woody material in the spring, especially in years of higher snow loads. We keep it picked up to reduce the fire hazard and to open microsites for shrub regeneration. Initially, we removed several prize bitterbrush that were located within the defensible open space surrounding the house. Now, however, a neighboring housing development, which completely scraped off the shrubland, is likely our most effective fire deterrent.

I made a winding dry creek bed connecting the aspen grove to a dry meadow located much closer to the house. At the aspen end of the dry creek bed I planted coyote willow, Woods rose, and western thimbleberry. The dry meadow is a drip-irrigated construct giving us a sense of verdant herbaceous space that transitions to the gray-green shrubs. To create the dry meadow, I collected seeds of favorite plants from nearby meadows. In addition, I also received many seeds from the Nevada Native Plant Society (then known as Northern Nevada Native Plant Society), which at that time organized seed collectors and seed cleaners and mailed out seed to members upon request, all for the price of postage. I simply broadcast-seeded and initially protected the area with chicken wire to keep out the cottontails and jackrabbits during establishment. Plants that have successfully established in the meadow include mountain goldenrod, meadow penstemon, sticky cinquefoil, nettleleaf horsemint, spiked sidalcea, Lewis monkeyflower, and Hooker evening primrose. I transplanted plugs of Douglas sedge from a trail along a nearby stream, and it has filled in the bottom part of the dry creek bed. But the most prolific meadow plant is the rhizomatous Nuttall sunflower—it is the glory of the yard in August, with its seven-foot-tall wands, and is the object of desire of sparrows and finches in the fall when in seed. I originally dug up several clumps of it from the garden of Margaret Williams (a founder of NNPS), and she told me to plant it where I wouldn't mind it taking over. Well, it has taken over the meadow, but its summer bouquets are worth it, and I manage it by digging it up at the margins to supply my friends' gardens.

The abundance of rocks onsite provided source material for low dry rock walls and a scattering of focal boulders. Now there are a number of showy natives growing near the rocks that were either planted from nursery stock or seeded. Early bloomers among these natives include yellow currant, wax currant, and little-leaf mountain mahogany, all of which attract many pollinators, especially native

solitary bees. I initially wanted some large rocks in the decomposed granite that were unadorned with flowers and rather zenlike; however, I reneged on my intentions after western columbine and cardinal monkeyflower self-established from their original placement and now attract hummingbirds for much of summer.

I planted several Jeffrey and ponderosa pines as individuals or in threesomes in the southeast, southwest, and northwest areas, for interest among the shrubs. Two Fremont cottonwoods and a black cottonwood grow on the north side of the property to provide shade and screening (from neighborhood houses) in summer and color in autumn. Two garden-grown western virgin's bowers are planted at the edge of the master bedroom deck. They annually create a 150-square-foot privacy screen—and that's with pruning. These plants were grown by Margaret Williams, and I am honored to have her plants on our property as a tribute to the best-known native plant garden enthusiast in the Reno–Sparks area in her time.

## Accidental Journey into Native Plant Landscaping

Randall Nish, Provo, Utah—*Foothill Zone*

My accidental journey into native plant landscaping began as a boy. I spent countless hours wandering the rolling prairies of southern Alberta, Canada, and the then-vacant foothills east of Provo, Utah. My parents, Dale and Norene Nish, introduced me to landscaping as we attempted to tame a difficult plot of ground around our new home in the Indian Hills subdivision of Provo. We used red sandstone gathered (with permission) from private land east of Heber City, Utah, to create a sculpted rock garden around the yard. As a child, I found the subject of western animals and birds to be approachable and familiar. However, the subject of native plants was cloaked in a confusing array of common names and unintelligible Latin.

Deer hunting in Hobble Creek Canyon, Utah, is an experience in close-quarters combat with plants. The steep terrain is covered with brush that tears at your clothing and scratches your skin. It was during one of these hunting trips in the early 1980s that it occurred to me that I could identify every mammal and bird in the area but could not name more than a couple of plants. I decided then that I could not claim to be an educated outdoorsman until I learned more about our native plants. I started a casual search for books about western plants in the gift shops of national park visitor centers and the local Utah State Extension Service. The books in the university and city libraries were useless to me because of their dense, technical language and the hated Latin words. The most helpful books were Ronald M. Lanner's *Trees of the Great Basin* and Hugh N. Mozingo's *Shrubs of the Great Basin*, along with a smattering of Utah State Extension Service booklets, including Sue Nordstrom's *Creating Landscapes for Wildlife*.

Two things happened that changed my casual interest in native plants to one of heightened urgency. First, our oldest daughter was born with cerebral palsy and would have only limited access to the natural world. Second, in 1989 we bought an empty lot in the Edgemont area of Provo, Utah, and decided to build a home. The idea was to bring the outdoors to our back yard, so our handicapped daughter could learn about her western heritage. The biggest obstacle was a lack of knowledge about the use of Utah native plants in residential landscapes. The resistance of our neighbors and my long-suffering bride to the then-unfamiliar topic of native plant landscaping was another problem. A compromise was reached where I could use native plants in the back and side yards while planting conventional Kentucky bluegrass and ornamental plants in the front yard. Our back yard became a decades-long living laboratory that continues to educate and fascinate.

Our home is located on the Provo Bench in northeast Provo, at an elevation of 4,770 feet. The property is 0.37 acres (1,497 $m^2$) and was originally the site of a fruit orchard and, later, a cow pasture. The topsoil consists of four to six inches of Bingham loam. The subsoil reaches ten to twenty inches deep, followed by a deposit of coarse sand, gravel, and river rock presumably hundreds of feet thick. Soil drainage is such that one can pour a five-gallon bucket of water onto the soil and it will disappear in less than a minute.

The first step in the back and side yards was to reduce the amount of lawn by creating large planters around the perimeter of the lot. Once the borders were established, we decided to plant trees for privacy and native fruit crops that would attract birds but could also be eaten by humans in an emergency. We also wanted different watering zones for three distinct plant communities. We relied on publications from the Colorado and Utah State Extension Services and occasional field trips to the local canyons and foothills to plan our landscape.

A breakthrough came when I heard a lecture at the newly created Utah County chapter of the Utah Native Plant Society, where the researcher taught that western landscapes are dominated by groves of small trees rather than large, single specimens. After this, we began to plant a large grove of Gambel oaks, sprouted from acorns gathered from a single surviving grove located next to Centennial Middle School. It is important to harvest the acorns from groves

at the same elevation where you intend to plant. It took us two years to work out a method for sprouting Gambel oak. The acorns were gathered in August and kept in a plastic jar filled with moist potting soil. The container was left in the refrigerator until the following February. Each sprouting acorn was transferred to a small pot and kept next to a south-facing window until it could be safely planted in the yard the following spring. We tried planting a few bigtooth maple and netleaf hackberry seedlings among the small oaks to simulate native Gambel oak groves in the area. The hackberries thrived, but the bigtooth maples had a 50 percent mortality rate. Creeping Oregon grape was used as an understory plant, due to its symbiotic relationship with Gambel oak. Examples of this plant community can be seen along the Provo River trail in Provo Canyon.

The southwest corner of the house was reserved for heat-loving trees, such as curlleaf mountain mahogany and Utah juniper. Attempts to grow singleleaf pinyon pine have been a complete failure. The trees were bought from a local nursery. Shrubs included green Mormon tea, winterfat, and sand, black, and big sagebrush. Perennials we used included Greek yarrow, scarlet globemallow, pineleaf penstemon, scarlet bugler penstemon, Palmer penstemon and scarlet hedgenettle. Scarlet globemallow was sown directly into the ground, using seed taken from an area east of US Highway 89 and south of Springville, Utah. The land was going to be bulldozed, so we also dug as many sego lily bulbs as we could use.

Over the next two years the sego lilies all died out, but the globemallow began to grow and continues to thrive almost twenty years later. The penstemons and scarlet hedgenettle are incredible hummingbird attractors, but the hedgenettle must be planted in areas where it will get water at least once a week.

The southeast corner of the home was planned as a bigtooth maple grove. This area is shaded in the morning by a large honey locust tree in the neighbor's yard. Unfortunately, the bigtooth maples have failed to thrive. The planter is slowly being turned into another Gambel oak grove by the scrub jays, who stash acorns in the bark mulch. When the scrub jays plant an oak in a good spot, we let it grow. Unwanted seedlings are removed with a shovel.

The remainder of the yard is devoted to edible shrubs, such as currants, gooseberry, elderberry, chokecherry, and Saskatoon serviceberry. The currants and gooseberries are generally unkempt looking and are of little interest to the birds. We planted elderberries and chokecherries in spite of being discouraged from doing so by local gardening experts. The elderberries are beautiful and are moderately useful to the birds. They require heavy pruning and thinning each year to keep them attractive. The chokecherry and serviceberry bushes are the star performers in our landscape. They reliably produce heavy crops of berries and are hugely popular with the native birds; they seem especially important for fledgling robins. The shrubs have deep green foliage all summer, with good spring and fall color. The chokecherries tend to sucker, and they will grow over twenty-five feet tall if given good soil and some summer water.

The biggest surprise about our yard has been how much we have enjoyed the experience of the last eighteen years. We have been amazed by how our yard has become a magnet for native birds. The Gambel oak grove is now twenty feet tall and is the preferred habitat for hummingbirds, robins, California quail, scrub jays, and blackheaded grosbeaks. Curiously, the non-native birds, such as starlings, avoid our yard. Another important lesson has been that all of our gardening problems are related to water. In the native plant areas, we find that we have few weeds and no pests. Weeding takes about an hour a month of light work. It is the water-intensive parts of the yard such as the bluegrass lawn and vegetable garden that require the most money and time.

If we were doing it over again, we would start out with even less lawn, both because of its high water requirements and because today's children spend little time outside. We would also hire a landscape architect experienced in Utah native plant landscapes. This would have saved us time and money. But it would also have deprived us of the wonderful experience of trial and error. Prepare the best you can. Take your first, best shot at a natural landscape. Don't worry about making a mistake, because nature will teach you how to sort things out if you are willing to learn.

## Tabula Rasa

Carl Dede and Angie Evenden, Boulder, Utah—*Semi-desert Zone*

In the mid-1990s, we bought twenty acres just outside of Boulder, Utah, with the vision of living there someday. The parcel, located at the edge of a valley at 6,300 feet elevation, supports a mosaic of Navajo sandstone slickrock and sand dunes, with native pinyon pine and juniper woodland. Most of the land was undisturbed and in good condition—however, four acres of deep sand along one edge of the property had been scraped clear of pinyon, juniper, and sagebrush a number of years before, presumably in preparation for a hay field that was never planted. This disturbed area, which included our eventual house site, was covered with annual weeds dominated by tumble mustard, tansy mustard, and Russian thistle, with only an occasional clump of native sand dropseed or Indian ricegrass. No sagebrush, pinyon, juniper, or other woody species remained on this portion of the property, although some beautiful old-growth trees grew along the edge. Being plant lovers, we saw the possibility of landscaping and restoring the disturbed area with native plants. Our vision was to broadly mimic the species and patterns of the naturally occurring native sagebrush and bunchgrass communities and to augment these with other suitable and attractive natives. We wanted to create a landscape that would be beautiful as well as drought resistant. In anticipation of our project, we began collecting seed of local native grasses, wildflowers, and shrubs and identifying other sources of plants for our project.

We built our house in the late 1990s, and that is when our restoration project began. After construction was completed in the fall, the house sat in the middle of three-quarters of an acre of completely empty sand, contoured and bare. Our first act of landscaping was to hand-drill in small caches of Indian ricegrass seeds, two inches deep, every eighteen inches or so, covering the entire bare area. That effort took four days on our hands and knees, sometimes in the rain. The damp conditions that fall and winter were perfect for producing the moist pre-chill needed for germination. The supply of our hand-collected grass seed was insufficient for a project of this size, so we augmented it by purchasing several pounds of Indian ricegrass seed from a local seed supplier. We were able to find seed collected within a hundred-mile radius of our property. However, it took three years before an abundant stand of ricegrass was established. We learned that as the delicate young blades emerged, they were promptly eaten by a hungry population of cottontail rabbits. Eventually we countered this by installing plastic vexar tubes (typically used to protect tree seedlings) over each small ricegrass clump and leaving them there until the plants gained sufficient

volume to outpace the rabbit consumption. Now, several years later, we have a self-perpetuating stand of ricegrass that does not need protection. That first winter, we also broadcast-seeded a variety of native grasses, wildflowers, and shrubs that we collected in the vicinity of our property. These seeds also took their time in germinating, and it would be a surprise each year to see what came up.

In the first spring after construction, we began the second phase of our landscaping project—starting with the installation of a drip irrigation system. During this phase we planted a number of shrubs and trees, mostly acquired in one-gallon containers from local Colorado Plateau native plant nurseries. Our approach for reestablishing sagebrush, however, was to transplant small plants from elsewhere on our property. For the next five years, each spring, and sometimes in the fall, we would plant dozens of shrubs and trees each season and hook them up to the drip irrigation system. We selected a combination of local natives that we knew would have occurred on the site, as well as other natives that would be well-adapted and attractive in our new landscape. Some of the shrubs that we planted are cliffrose, littleleaf and alderleaf mountain mahogany, oakleaf sumac, Apache plume, big sagebrush, sand sagebrush, sand penstemon, fourwing saltbush, rubber rabbitbrush, fernbush, roundleaf buffaloberry, Utah serviceberry, and green Mormon tea. The variable foliage color and texture of these species provide a lot of visual interest. We were also able to purchase larger pinyon pine, Utah juniper, and ponderosa pine (four to six feet tall) from a local nursery for fall planting, giving us a head start on tree structure on the site. Because we have access to irrigation water from nearby Boulder Mountain, we also opted to plant a few fast-growing Fremont cottonwoods. For spots of color, we planted the brilliant magenta flowered desert four o'clock; datura, with its large, night-blooming,

tubular white flowers; yucca; and a variety of penstemons. In a shady corner by the deck, we fenced rabbits out of a small area for a blue grama lawn, presided over by alcove columbine, echinacea, Hooker evening primrose, coral bells, and blanketflower.

Due to the proximity of an alfalfa field, large numbers of mule deer browse their way across our property at night. In addition, there were occasional visits from stray cows, llamas, and even elk. It quickly became evident that most new shrubs and trees required protection (at least at first, when they were small). We made four- and five-foot-high cages from welded wire obtained from the farm supply store, or field fence reclaimed from the local dump, to protect plants and trees from browsing, trampling, and rubbing damage. Each cage was anchored to the sand by eight-inch-long, U-shaped wire stakes. As the plants prospered and outgrew them, the cages could be moved to smaller plants. Some species, gaining sufficient size, no longer needed protection from deer; these included sagebrush, rabbitbrush, roundleaf buffaloberry, sumac, and Apache plume.

With the deep, loose sand on our site, small plants needed frequent watering to become established. The drip system provided this crucial early moisture, and battery-operated timers made it automatic. Some area sprinkling was done to help establish the bunchgrass at first—although this practice also encouraged the weeds. As some plants grew up, their emitters could be removed. As the plants around the house took off, they reproduced prolifically (especially sagebrush), resulting in numerous seedlings available for outplanting. In this way, we quickly had our own nursery and were able to transplant these seedlings in other areas, as well as sharing plants and seeds with others in the community.

Dealing with weeds has been an integral part of our project. We quickly learned that the "bare" sand was not entirely barren, but was well stocked with a seed bank of undesirable weeds as well as desirable natives. From the start of our project, we aggressively weeded the restoration area around the house. Fortunately, the weeding task has diminished each year as the native perennials became established and spread. We were very pleased that several native wildflowers and grasses volunteered, adding significant color and beauty to our landscaping project. These species included sand dropseed, pale evening primrose, purple aster, sunflower, and hairy golden aster. The latter species is a favorite fall food of the lesser goldfinch.

We succeeded beyond our wildest dreams and were amazed at how quickly we were able to establish a flourishing and beautiful native landscape. The combination of sand substrate and ample water resulted in ideal growing conditions and unbelievably rapid growth for the native shrubs. Littleleaf mountain mahogany planted at ten inches tall, with a few stems, grew within five years to a size five feet in height and four feet in circumference. Fourwing saltbush achieved heights and diameters of six feet in each direction within the same timeframe. Similarly, almost all of the shrub species we planted shared these dramatic gains in size.

Our project continues to evolve as the years go by. We now find ourselves removing plants from the landscape area around the house to maintain more openness. Since this area now needs little attention, we have shifted our efforts to restoring the remaining disturbed acres by planting grasses and sagebrush. Creating our native landscape has been an extremely gratifying undertaking, and we enjoy its beauty each day. Our project has attracted the interest of many local community members who are interested in establishing similar native landscapes, and our site is regularly visited by people for ideas and inspiration.

## Starting Small, Becoming Emboldened

Ann DeBolt and Roger Rosentreter, Boise, Idaho—*Semi-desert Zone*

In 1990, as relatively new homeowners who also happened to be agency botanists, we decided to embark on establishing a small native garden in a non-landscaped portion of our yard. When we purchased this first home in 1988, it became apparent that this portion of the property had been used for parking cars and had received little attention, in spite of the fact that it served as the entry area into the back yard and to the back door of the house. Consequently, the silty clay loam soil was compacted and occupied by weedy, non-native species such as prostrate knotweed (*Polygonum aviculare*), bur buttercup (*Ranunculus testiculatus*), spurge (*Euphorbia glyptosperma*), common mallow (*Malva neglecta*), and prostrate pigweed (*Amaranthus blitoides*). Because the area would continue to serve as an access corridor to the back portion of the house and yard, it was necessary to create pathways through it. And rather than rototill or churn up the entire 30 × 30 foot area and encourage additional weed seed germination, we decided to hand-pull the undesirable species in the spring, when the ground was moist, and plant around the edges of the area, leaving much of the middle open for walking. We used pea gravel as mulch for the unplanted areas and as our walking surface, because it was readily available, affordable, and would provide good drainage for our compacted site and the type of plants we wanted to grow.

Our project area was on the east side of the house and adjacent to the street. Unlike the rest of the yard, it had no automatic irrigation system. There was a silver maple to the north that did cast a small amount of shade onto a portion of the area for several hours of the day, but the site was generally open, hot, and sunny. It was the perfect place to start small with our native landscape, as this particular portion of the yard could only be improved.

At around this same time, Ann began coordinating a native plant sale for the Idaho Native Plant Society, and through the process of acquiring plants for the sale from native plant nurseries and the yards of fellow plant enthusiasts, we had access to a variety of species. While we wanted to emphasize native species in this garden, water conservation was also an objective, so we broadened our palette to include plants native to other regions of the country, in addition to several drought-tolerant cultivars.

Idaho natives:

*Achillea millefolium* (western yarrow)
*Erigeron pumilus* (shaggy fleabane)
*Eriophyllum lanatum* (Oregon sunshine, woolly sunflower)
*Lewisia rediviva* (bitterroot)

*Linum lewisii* (Lewis flax)
*Opuntia polyacantha* (Starvation prickly-pear cactus)
*Penstemon deustus* (hotrock penstemon)
*Phlox hoodii* (Hood phlox)
*Salvia dorrii* (desert sage)

Native to the United States but not to Idaho, or as otherwise indicated:

*Echinacea purpurea* (purple coneflower)
*Festuca ovina* Glauca' (blue fescue; ornamental cultivar)
*Penstemon eatonii* (firecracker penstemon)
*Phlox subulata* (low-growing phlox; ornamental)

The area was planted in stages, as species were acquired over a year-long period. We did not amend the planting holes (as is often recommended), but assumed that the pea gravel mulch throughout the open areas would work its way into the soil to improve drainage as time passed. Once planted, maintenance of this site required little effort. We irrigated those plants that seemed to need it, or would flower longer if they received water, once a week by hand. This included purple coneflower, Lewis flax, and the ornamental phlox. Others were watered approximately every other week, though we tried to withhold water from the pricklypear cactus and from the bitterroot once it became dormant in the summer. Other regular but infrequent maintenance included the following:

1. hand-weeding undesirable species that germinated in the gravel mulch
2. cutting back the Lewis flax after it flowered in the spring, thus reducing the seed production and spread of this species and promoting reflowering
3. removal of leaf litter that blew or fell into the area, as leaf litter promotes rotting during the winter if left on top of plants

4. cutting back the purple coneflower and firecracker penstemon flower stalks in late winter/early spring

We used local rocks to protect and highlight our plantings, and the area flourished and provided color and texture for us and passing motorists and pedestrians throughout most of the year.

Emboldened by our first success at incorporating native plants into our landscape, we decided to take on other sections of the yard. Modifying our landscape in sections worked perfectly for us, as we were on a corner and had built-in breaks in the existing landscape, such as a driveway, a sidewalk, and a concrete walkway. Our side yard was the next phase, with a portion of the front yard as phase three.

Both the side and front yard had a Kentucky bluegrass lawn, with shrubs and trees restricted to the edges, and an existing irrigation system. While there were several options for removing the lawn—rent a sod cutter, suffocate the grass with cardboard or newspaper followed by mulch to allow it to die in place, apply glyphosate (Round-Up) herbicide—we chose to use Round-Up herbicide in early summer and to withhold irrigation, so that by fall the lawn would be dead and ready for planting. This approach worked well for us and eliminated all risk of damaging the irrigation system, which we wanted to leave in place and functional for us as well as future homeowners.

The side yard (phase two) measured approximately twenty feet wide by seventy-five feet long. Here is a list of some of the species planted directly into the dead sod.

Idaho natives:

*Artemisia cana* (silver sage)
*Eriogonum heracleoides* (Hercules buckwheat)
*Eriogonum umbellatum* (sulfurflower buckwheat)
*Purshia tridentata* (bitterbrush)
*Townsendia florifer* (dwarf aster)

Western United States natives:

*Cercocarpus montanus* (alderleaf mountain mahogany)
*Datura sp.* (sacred datura)
*Penstemon palmeri* (Palmer penstemon)
*Penstemon pinifolius* (pineleaf penstemon)

The plants came from a variety of sources, including native plant nurseries and as seedlings, from the yards of various other botanists. The area was mulched with large pine bark chips, which, once weathered, gave an appearance of rock mulch without the weightiness and permanence of rock. This area was irrigated by hand on an as-needed basis, which meant once every second or third week.

All the plants thrived except for pineleaf penstemon, which survived but would have flowered much more prolifically with weekly irrigation. Silver sage, which can spread by root sprouts, requires annual late-winter pruning to keep it from becoming too large and woody. Should it receive too much water, root sprouts can become a problem.

We chose to keep phase three—the 25 x 25 foot, north-facing front yard—simple, planting it to a ground cover of kinnikinnick (*Arctostaphylos uva-ursi*), with an intermittent scattering of non-native bulbs (crocus, hyacinth, miniature daffodils) for spring color. Kinnikinnick plants were placed at random with approximately two feet between them, knowing they would spread up to three feet wide, but that they would do so slowly. In the hot, dry, alkaline Treasure Valley where Boise is located, kinnikinnick grows best if it is given a moderated environment with partial shade or a north-facing aspect. It also requires twice-weekly irrigation in the heat of the summer, at least until the second year, when it is well established. This area was also mulched with pine bark, but mulch was kept away from the base of the plants to prevent rotting. This is generally required for most native plant species, particularly those with a herbaceous growth form.

We have since moved to a much larger property (1.8 acres) where we have planted a combination of traditional and native gardens. We continue to use many of the same species as listed above, but have become fond of a number of others for their ability to thrive and to beautify Treasure Valley landscapes: curlleaf mountain mahogany (*Cercocarpus ledifolius*), prairie smoke (*Geum triflorum*), sticky geranium (*Geranium viscosissimum*), Indian ricegrass (*Achnatherum hymenoides*), fernbush (*Chamaebatiaria millefolium*), shrubby penstemon (*Penstemon fruticosus*), maple mallow (*Iliamna rivularis*), and Munro globemallow (*Sphaeralcea munroana*). All are commercially available and are lovely examples of what our native habitats have to offer to those who wish to have a little bit of nature in their own back yard.

## Learning by Trial and Error, or Life Is Change

Roger Kjelgren, Logan, Utah—*Foothill Zone*

I actually started with native plants in Illinois, where intermountain plants would be exotic, and hardwood forests native. I found Illinois natives much more interesting than your standard, run-of-the-mill landscape plants. Moving to Utah sixteen years ago, I was a bit taken aback with the apparently limited selection of plants to choose from. Very few trees, lots of strange shrubs, and all these funny perennial species, including so many DYC's (damn yellow composites) that looked like mutant dandelions.

However, over time and with lots of hikes around the state, where I got up close and personal with the plants around the Intermountain West, my attitude changed. I began to appreciate how tough some of these plants are, growing out of rock with no water and no visible sources of fertility, how attractive so many are, and, particularly for low, slow-growing shrubs, how low-maintenance they are. The other, and more subtle, revelation was the sense of space that allowed me to see each plant as unique, something that is hard to do in the wetter parts of the country, where the native vegetation is usually a wall of green.

I also began to appreciate the diversity of color in the intermountain native perennials and the nice small size of many of the perennials and shrubs. Small size and tight growth meant less maintenance, which I really liked, because I'm not into recreational pruning. Finally, I began to realize that I really liked low-water-use landscaping, after dragging hoses to irrigate turf on our sloped front yard and watching the water run off onto the sidewalk. So, since life is change, it was time for a change to native plants.

I started with ripping out all bluegrass turf in my front yard, putting in rocks to maintain grade where the ground had sloped down next to the sidewalk. I then made two small patches of turf, in this case a warm-season, shortgrass prairie mixture of blue grama, sideoats grama, little bluestem, and buffalograss. The grass was surrounded by beds containing a mixture of non-native and native perennials and shrubs. I really learned what drought-intolerant indicator plants looked like, as I planted a lot of coneflowers that would wilt after several days without water, while all the other plants were fine. Right next to the coneflowers I planted cliffrose, which taught me about native plants that are intolerant of water, as it got the same frequent watering as the coneflowers, and died over a period of two years.

The biggest lesson from my willy-nilly mixture of natives and non-natives was the importance of space, a trademark feature of native plant landscapes. My landscape looked like a jumbled mess; without space between the plants, it was

impossible to make sense of the landscape. I knew something was amiss but couldn't put my finger on it, until I asked my wife if she liked our native plant landscape, and she said flat out, "No, it's a jumbled mess"—the first clue. So I pulled out all the plants that were even a little bit intolerant of drought, including the small prairie planting, because it was too small to look like anything other than an overgrown lawn.

To make the landscape more legible, I broke up the entire front yard with paths so no one bed was larger than about ten feet. Lo and behold, it worked, as my wife gave it her seal of semi-approval (the plants still looked a bit ragged, in her opinion), and I liked it better. I got rid of the smaller perennials and put in lots of penstemons (mostly *Penstemon strictus*, or Rocky Mountain penstemon), buckwheat, and evening primrose. I replaced the prairie with a native ground cover I was intrigued with, trailing daisy (*Erigeron flagellaris*), to see if it could be made to look like a lawn. I also set up an extensive drip irrigation system, using lots of spaghetti tubing where every plant had an emitter.

This new setup worked much better, in the sense that things didn't need much water, I had lots of color in the spring, and I had two marvelous lawns of trailing daisy that turned snow white for several weeks in May. However, this setup also started to fray around the edges after a few years. Putting drip emitters at each perennial proved to be totally impractical, as many of them, particularly

the penstemons, are short-lived if they are overwatered—where they died I was irrigating empty ground. I also discovered that a large number of native perennials needed to be cut back (deadheaded) in order to look good, which was a lot of work, in my estimation. Finally, the trailing daisy was a bit more complicated than I realized. It was dense enough to be fairly competitive against many weeds—but those it couldn't outcompete included many volunteer grasses from the former prairie and or some broadleaf weeds. Because both trailing daisy and the weeds are broadleaf perennials, there are no herbicides to control only the weeds. The only broadleaf weed to actually be a problem was field bindweed. A spot application of glyphosate and 2,4-D took care of that. I transplanted a few trailing daisy plants and watered them, and their phenomenal growth rate filled in the blank spot in a hurry. Another thing I learned about trailing daisy was that the runners can become so dense at the end of a season that they can shade out the leaves, so the plant cover thins out and lets in weeds. While there is a bit more to learn about maintaining a trailing daisy lawn, it does appear that if I mow the runners in late summer, the leaves come out stronger the following spring.

After a few more years of pondering this arrangement, my current state of mind is to simplify my native landscape by going with one or two long-lived perennials that need minimal cutting back—such as pink smoke buckwheat (*Eriogonum racemosum*) and lacy buckwheatbrush (*E. corymbosum*)—and keep shrubs such as manzanita (*Arctostaphylos* hybrids), littleleaf mountain mahogany (*Cercocarpus intricatus*), and Utah holly (*Mahonia fremontii*). These are very drought-tolerant and need little or no supplemental water, require no pruning, and generally look good. We'll see what happens next.

# Designing on the Native Plant Frontier

Bettina Schultz, Elk Ridge, Utah—*Foothill Zone*

## Prologue

I garden with Susan Meyer on a one-acre property in Elk Ridge, Utah, a few miles south of Provo. When we moved into the house in 1989, the front yard was a wasteland of ratty lawn, on a north-facing slope in the foothill zone. There were some irises and a few trees and shrubs next to the house, a small pine tree in the northwest corner, and lots of whitetop in the grass. The lawn area closest to the house was the most fertile and got the most shade. Then there was a fairly steep slope down to another flattish area at the bottom of the yard. Our soil was full of clay, but since there were plenty of rocks in it and we were on a hill, there was no problem with drainage.

Over the next twelve years, we reduced the lawn area by clearing out the edges and planting shrubs, trees, bulbs, and mountain wildflowers, and installing a drip system to cut down on the area that needed to be sprinkled. I was the principal lawn-mowing person, and I quickly tired of hauling the noisy gas mower up the steep part of the lawn. So we took the grass out of the steepest part first, replacing it with some big stepping stones. Then we cut a big circle out of the center of the steep area and planted it with flowers. I built an informal rock retaining wall above and below the circle. Eventually we cut away most of the rest of the steep part, extending the retaining wall and planting penstemons along the top and bottom. By this time we had removed about half the grass.

## Taking the Plunge

In 2002, we decided to finally go all the way and kill the bluegrass. By this time I had some experience designing Utah Heritage public demonstration gardens, but I discovered that letting go of our very own lawn was a scary prospect. I mulled over the idea for a long time before we actually did the deed. In fact, there were several ideas that played into the eventual design.

I was stuck on the notion that we needed something low and flat to replace the lawn. Somehow, ground covers didn't feel like the answer. Then one day I saw a little photo on the Internet, a small area, full of shrubs of varying heights, colors, and textures, with a walkway next to the planting. A light bulb went on in my head. While I still needed some low, flat areas to set off taller vegetation, I could use hardscape—walkways. That idea opened the door for the rest of the design process. The thing about lawn, after all, is that you can get to wherever you need to go in the yard by walking across the grass. So I started looking at the

yard in terms of destinations—where did my pathways need to go?

I measured the area and made a simple map of the existing features I knew we wanted to keep: the trees and shrubs around the edges, the circular bed, and the retaining wall with its penstemons. I made lots of copies of the base map, so I could scribble on one with abandon and always have another copy to try again. Then, by thinking about where we typically wanted to go in the yard (and actually walking around in it), I was able to plan the path system. The path system turned out to have another advantage: it broke up the vast area of the front yard into smaller chunks, which were easier to think about. I made the main paths wide enough to accommodate our garden cart, knowing that as the plants grow up, the paths will become less noticeable.

I started to think about what the yard would look like from two main vantage points: from the driveway, as you approached the house, and from the living room windows that looked out on the front yard. When I thought about shrubs, I remembered something I had seen years before. We were driving through central Utah, and I looked off across a wide valley to the low hills beyond. The floor of the valley was covered with blue-green sagebrush, and as the land rose up the hillside, there were bands of different colors of vegetation—olive green, gray-green, pale gold, and the dark blackish green of distant junipers. The bands were fairly distinct, but the range of colors was subtle, and it gave me a feeling of peace. I decided to use this idea in the yard by making bands of different-colored shrubs. Since we already had a band of lavender and another of day lilies across the east side of the yard next to the driveway, I just repeated the bands, using shadscale, winterfat, spiny hopsage, lacy buckwheatbrush, green Mormon tea, and black sagebrush. Desert sage and Indian ricegrass filled the area between the main path and the existing bulb bed along the front walk, with another section going to low rabbitbrush and more Indian ricegrass.

Then there was the circle bed. I decided to put a spiral path through the circle and fill the area around it with the lowest-growing flowers. It was Susan's idea to substitute sand for our clay soil in the spiral planting area. She hauled many loads of dirt away, and many loads of sand in, to make that happen. The flowers in the spiral were planted in a random mixed pattern, like a wildflower

meadow. They have been extremely happy with the drainage and low fertility.

I put some of the larger flowering plants in the areas adjacent to the spiral bed, keeping them grouped together as blocks of color. There were a few accent plants at focal points in the design: an alkali sacaton grass "fountain" at the tail end of the shrub area, an Apache plume in the flower area just east of the spiral, and some blue grama grass among the flowers around the outside of the spiral bed. Since the view from the driveway is toward the west, the grasses look great when backlit by the afternoon sun.

In the shadier, more fertile, drip-irrigated area closer to the house, I put plants that could take more water: a bank of firechalice, a row of little bluestem grasses, and a special spot for "foreign" penstemons—not native to Utah, but gifts from various friends. We also added some shrubs and another row of firechalice in the drip irrigation zone around the edges of the yard.

## An Ongoing Process

Our yard is an experiment in progress. Many of these plants had not been used before in a home landscape, in a foothill setting. Could the desert shrubs handle the snow load? Which plants would be the winners and which the losers? What would the maintenance be like? We've been learning from the yard over the past several years since we planted it.

So far, the shrubs have not had any real trouble with the snow or the amount of precipitation in our zone. Spiny hopsage turned out to be a mistake. It loses its leaves by late summer, so we interplanted a stand of "pink smoke" buckwheat, which blooms when the hopsage goes dormant. We'll see how it works. The low rabbitbrush got too floppy in our soil, so we replaced it with a group of yellow-flowering lacy buckwheatbrush plants—that species has been very satisfying. Since we left the dead sod from the former lawn in place after killing it, some plants got surprisingly large because of the high fertility. We're hoping that this effect will diminish over time.

Some of the real winners have been little beebalm, silver buckwheat, showy sandwort, desert sage, fragrant evening primrose, and all the grasses. An unexpected bonus has been the fascinating array of pollinators that visit our flowers, from native bees to butterflies, moths, bee-flies, and hummingbirds. As for maintenance, it's been surprisingly easy. In the spring we get a carpet of seedlings around the shrubs. Weeding them out takes the two of us a few hours. We hand-water new plants through their first summer, and haven't watered at all since then, except for the little beebalm and the areas on drip. Watering with drip irrigation involves turning the faucet handle on about an eighth of an inch in the morning and closing it several hours later. As the shrubs grow up, they need a bit of pruning to keep them from getting gangly. In the spring, I whack the bunchgrasses back to a few inches high, to tidy them up and let the new growth show itself. I definitely don't miss the lawnmower.

*Littlecup penstemon*

# The Plant Palette

Chapter Seven

This chapter contains the specific information you will need to choose the plants that will populate your native landscape. The species we have included in the Plant Palette were chosen from hundreds of native candidate species based on several criteria. First, the plant had to be attractive, if not astonishingly beautiful. This, of course, is somewhat a matter of opinion, and the list adopted here is the result of working and reworking by several knowledgeable people with different tastes. Second, the plant had to be relatively quick and easy to grow in container culture in a nursery setting. We avoided certain favorites, like sego lily, that have proven slow and difficult to produce. Work continues on many of these hard-to-grow plants, and the time may come when they will be commercially available. For now, we concentrated on plants that are either already available or could be brought on line quickly if warranted by demand. And lastly, the plant had to be at least somewhat tolerant of the abuses that are frequently encountered in residential landscapes. Too much water, too much mulch, too much fertility, and too much competition from other plants are some common forms of abuse. Not all of the plants we included are entirely foolproof in this regard, but, by using the information provided for each plant, you should be able to create favorable conditions in your landscape for even the more finicky species. We narrowed down the list of plants covered in the Plant Palette to one hundred species that we consider to be the core species for creating regionally distinctive landscapes in the Intermountain West. Many more species could have been included, and it is perfectly fine to use species not included in this book in your native plant landscapes. Just get the information you need to meet plant cultural requirements (water, soil, light, and cold hardiness) and make sure that the plants you select really are native to the intermountain area. "Native" is a somewhat slippery concept, in that plants can be native to a very restricted area, a state, a region, a country, or a continent, and a few plants are naturally cosmopolitan (worldwide) in distribution. But just because a plant grows wild in a region does not mean that it is native to the region. Many species native to other places have been deliberately or accidentally introduced into the wild plant communities of our region. If you have any questions about whether a particular plant is native to the region, a good Internet resource is: plants.usda.gov.

Native plant species vary in the range of habitats that they can successfully occupy. There are many common plant species in the Intermountain West that occur over an amazing range. In some cases, the plant species may be made up of a series of races that are adapted to particular environments—these are called ecotypes. Ideally, it is the responsibility of the nurseries supplying native plants for horticultural use to make sure that the ecotypes being sold in a region are adapted to that region. This is particularly important with regard to cold hardiness. For example, desert sage plants from warm desert populations in the Mojave Desert have been found to be cold hardy only to Zone 7 (average winter minimum temperature from 0° to 10° F), while populations from cold Great Basin valleys are hardy to at least Zone 5 (average winter minimum temperature –10° to –20° F). These differences in cold hardiness clearly have a genetic basis. Similarly, it is quite possible that an ecotype of a widely distributed species that is from an area of higher rainfall will have a somewhat higher water requirement than an ecotype of the same species from a lower rainfall area.

There is another process involved in the apparently wide tolerance of many intermountain species. The plant may occur across a wide range of elevations, but in specific microhabitats, so that the microclimates are more similar across sites than the elevation range would imply. Firecracker penstemon is a case in point. This plant occurs from the fringes of the Mojave Desert up to rocky slopes at twelve thousand feet in elevation. At the low end of its range, it is usually found on shallow, sandy soils in sheltered canyons with partial shade. At the high end, it is again found on shallow, rocky soils, but in full sun as part of a low, perennial plant community. At middle elevations, firecracker penstemon is almost never found as a part of intact woodland or forest vegetation. Like many penstemons, it seems to be a road-cut specialist. It thrives on natural or manmade disturbances characterized by steep slopes, relatively high light intensity, and soils that are little more than rock debris. In these habitats it can escape competition and shade from other plants, as well as the pathogens found in richer soil under intact vegetation. The common elements in the cultural requirements for this plant include excellent drainage, soils very low in organic matter, steep slopes, and minimal competition. And even though its range includes wide variations in terms of precipitation, it tends to be found on the driest microsites available, especially at the upper end of its elevation range.

Many native plants were virtually unknown in the trade only a few years ago, and most designers, contractors, and homeowners have limited experience with them. In this guide to species for Intermountain regional landscapes, we try to give you enough information about each plant to incorporate it into an intelligent and workable design. This includes information on both aesthetic characteristics and cultural requirements. The Plant Palette entry for each plant is comprised of several kinds of information, including the common and scientific name of the plant, a habit illustration, a color photo, and a short description that highlights uses in the landscape and special features. Each Plant Palette entry also includes a series of icons that indicate plant characteristics, such as growth rate and flowering season, as well as cultural requirements. The Plant Palette is organized first into two groups, woody plants and perennials. Within each group, the plants are grouped by their water needs. Within each water-need group, the plants are listed in ascending order according to their average height at maturity. To facilitate design work, we also provide supplemental tables for woody plants and perennials. Plant features such as height and diameter at maturity are listed, along with cultural requirements and the page number for the Plant Palette entry for that plant.

In order to use the Plant Palette to find specific plants, we provide a comprehensive index to both common and scientific names at the back of the book. This includes all the commonly encountered variants for the common and scientific names of the plant, as well as the names used in the Plant Palette itself. This should make it possible to quickly determine if a plant of interest is included in the Plant Palette, and, if it is, to find the page number of its entry.

# Plant Characteristics

## Habit Illustration

The habit illustration for each Plant Palette entry shows the size and form at maturity of a typical member of the species. A scale figure is included in each illustration—this person is six feet tall. For species two feet tall or less at maturity, we just show the knees and ankles of the figure, scaled to be two feet tall, in order to provide a more detailed representation of the plant habit.

Many factors interact to affect the mature size of a woody plant in the landscape. These include both genetic variation within the species and differences in growing conditions. We chose to represent average rather than maximum mature height, so be prepared for the possibility of a somewhat larger plant over the longer term.

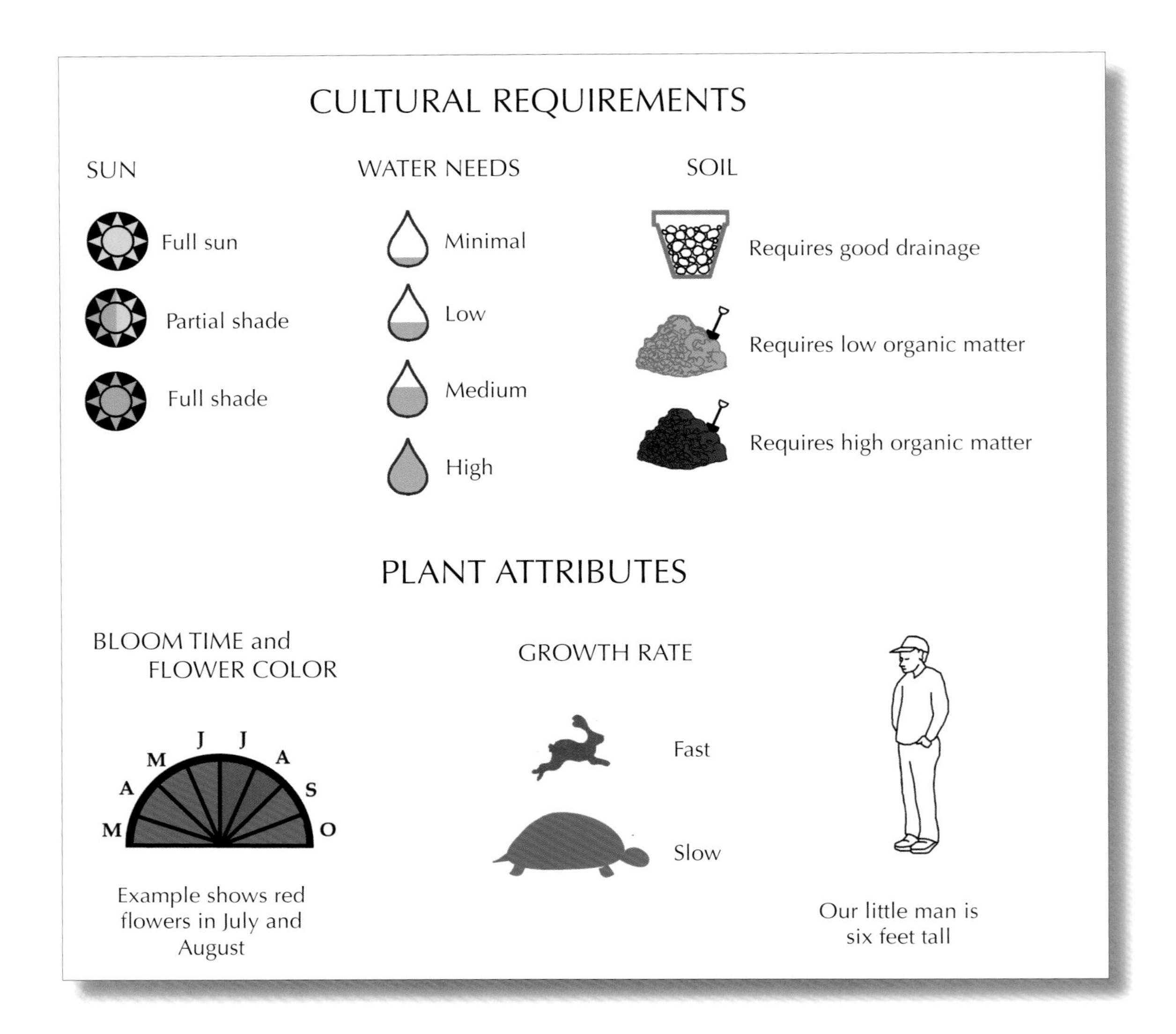


## Growth Rate

We include an icon to represent growth rate for woody plants that are either exceptionally fast growing (hare icon) or exceptionally slow growing (tortoise icon). Species that have no icon are considered to have medium growth rates. Growth rate, like maximum size, is variable, depending on growing conditions. Our estimates of growth rate are based on conditions specified in the accompanying cultural requirement icons. For example, if bigtooth maple is grown under the water regime for plants with a medium water requirement, it has a moderate growth rate (i.e., no growth rate icon). Under a high water regime, in an environment with little competition from other plants, it can grow more quickly. We do not include a growth rate icon for perennials because, in general, all of these plants have a fast growth rate in a landscape setting. Most can be expected to reach flowering size in two years, even from seed, and the increase in size from the first spring to the second can be phenomenal.

### Flowering Time

We include flowering time icons for all species whose flowers are large and attractive enough to represent a desirable feature of the plant. This icon is a semicircle divided into eight pie wedges, each wedge representing a month of the growing season, from March through October. For those months when you will find this particular plant in flower, the wedges are filled in with the color of the blossoms, while the remaining wedges are colored green. This gives you a sense of the flowering season at a glance. We use flowering times as observed in gardens in the semi-desert precipitation zone for the icons, not flowering times in the wild. These can sometimes be quite different, for example, for high elevation penstemons, which can flower in mid-spring in the valleys, when their native sites would still be deep in snow.

## Cultural Requirements

### Water Needs

Water needs for each plant are represented in icon form by a water droplet filled to four different levels, from almost empty, representing plants with minimal water needs, to full, representing plants with high water needs. These levels correspond to the amount of naturally available water in each of the four climate zones described earlier, namely the desert, semi-desert, foothill, and mountain zones, and also correspond to the water zones already designated on your landscape plan during the design process.

The water needs icons given for plants in the Plant Palette are intended to be conservative, in that they are based on the driest environment where the species grows naturally without the benefit of favorable microsites. The plants will generally be just as happy in a water zone that is one step higher than the one shown. Thus minimal- and low-water-use plants can be grown together in a low water zone, low- and medium-water-use plants can be grown together in a medium water zone, and medium- and high-water-use plants can be grown together in a high water zone. Plants can also sometimes be grown successfully within a water zone that is one step lower than the one shown in the icon if they are placed in a favorable microhabitat, such as a microhabitat that receives harvested water.

### Light Needs

Most of the plants in the Plant Palette fall into three groups in terms of light needs, those that require full sun, those that tolerate either full sun or partial shade, and those that prefer partial shade. By partial shade we mean either full shade for several hours during the day or dappled shade. In addition, there are a few native intermountain species that can tolerate continuous full shade. We indicate the range of light conditions that a species can tolerate using combinations of three symbols, a full sun symbol, a partial shade symbol, and a full shade symbol.

## Soil Needs

In general, all of the plants listed in the Plant Palette require at least adequate drainage, as described in the site analysis section. There are, however, many native plants, especially low- and minimal-water-use plants, that require exceptionally good drainage in order to thrive. These plants generally do better if the soil is sandy or gravelly, if the subsoil is cobbly, or if the site has enough slope to offer natural drainage away from the root system. We designate these plants with an icon that shows a flower pot full of coarse soil.

Species in the Plant Palette that have an aversion to soils with high organic matter are designated with an icon showing a light-colored pile of soil. Most of these are desert and semi-desert plants. Soils for these plants should never receive organic amendments or organic surface mulch. Gravel mulches, which drain water away from the plant crowns while still functioning to conserve subsurface water, will give far better results. Plants from the mountains that benefit from the higher fertility and water-holding capacity associated with organic matter are designated with an icon showing a dark-colored pile of soil. These plants may be grown in soil with organic amendments and organic surface mulches. Plants that have no soil organic matter icon are generally tolerant of the organic matter in ordinary topsoil but will not benefit from organic amendments except in truly heavy soils.

## Cold Hardiness

We do not include specific information on cold hardiness for species in the Plant Palette because all listed species meet our minimum cold-hardiness requirement. A primary criterion for inclusion in the Plant Palette was that the species needed to be cold hardy to USDA Plant Hardiness Zone 5 (average winter minimum temperature from –10° to –20° F). Most of the urban areas of the Intermountain West are in Zone 6 (average winter minimum temperature from 0° to –10° F) or warmer, but a few cities and towns in mountain valleys, such as Logan, Utah, are in Zone 5. Many of the plants in the Plant Palette are known to occur in nature in places at least as cold as Zone 5. A few do not grow naturally in places as cold as Zone 5 or even Zone 6, but they are successful in cultivation at much colder sites. For example, Apache plume does not naturally occur in places colder than Zone 7 (average winter minimum temperature from 0° to 10°

F), but it can be successfully cultivated at sites as cold as Zone 3 (average winter minimum temperature from –40° to –30° F). Natural distributional ranges therefore give a conservative estimate of cold hardiness. Species in the Plant Palette that are not found in nature in places as cold as Zone 5 are included based on their proven ability to survive long term in Logan, Utah.

## Interactions

Because of the interconnections between climate, topography, and soil, as well as the feedback from the plant community that develops under a particular set of conditions, there is generally a tight relationship between the water requirement for a plant and its light and soil requirements. For example, desert plant communities are generally characterized by high light intensity and low soil organic matter as well as low rainfall, so you are unlikely to encounter a minimal-water-use plant that thrives in shade or in very rich soils. Similarly, plants of mountain streamside communities generally grow in rich soils in partial shade, so high water use is often coupled with tolerance for shade and high organic matter. Thus the same combination of icons is often repeated for different plants in the Plant Palette. Plants with similar icon sets can be thought of as members of the same plant community, and can successfully be planted together.

## Cushion Globemallow

***(Sphaeralcea caespitosa)***

In the wild, this pretty little plant is restricted to the dry deserts of west-central Utah. It combines quite large, fragrant orange flowers with a diminutive stature and thick, silver-green leaves with scalloped edges. It can be grown in a variety of soils, but it will live longer if the soil is rocky and lean and watering is kept to a minimum. It will volunteer from seed, however, so you will likely have a persistent planting, even if individuals only live two or three years. Cushion globemallow is an ideal plant for a rock garden or low perennial border. It combines well with other petite desert plants like sundancer daisy and silver buckwheat.

*Special Features*: Like all its globemallow relatives, cushion globemallow readily forms hybrids with related species, so if you want the volunteers to look like their parents, be sure to plant only this species.

## Fragrant Evening Primrose

***(Oenothera caespitosa)***

Large, heavily perfumed flowers that open in late afternoon and bloom all night give this plant its name. Its rather large, deep green leaves and extravagant blossoms belie its tough, drought-hardy nature. Each flower blossoms only once, wilting and turning pink in late morning, but the profusion of new blooms lasts for weeks. The flowers feature abundant nectar that attracts hawk moths, and the anthers and cross-shaped stigma are held well forward from the petals, where the hawk moth cannot help but contact the sticky pollen. Be careful when sniffing the flowers, or a dust of cobwebby golden pollen will be left on your nose. This plant combines especially well with Utah ladyfinger milkvetch, and they are often found growing together along gravelly road shoulders.

*Special Features*: The sight of this plant flowering in the moonlight is unforgettable. Be sure to plant it where you will be able to enjoy the heady fragrance.

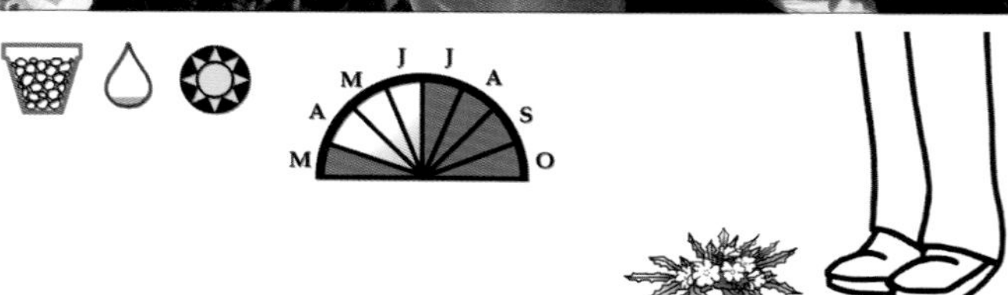

## Indian Paintbrush

***(Castilleja angustifolia* var. *dubia)***

Indian paintbrush is probably the best-known wildflower of the Intermountain West, yet it is rarely seen in gardens. The main reason is that this plant is dependent on the roots of other plants to help it obtain food and water—it needs a buddy to prosper, whether in the wild or in a garden setting. When you buy an Indian paintbrush plant in the nursery, it should already be potted up with a companion, so the roots have a chance to connect before being planted in the ground. We like to use big sagebrush as a companion plant for Indian paintbrush—the color contrast is beautiful. There are many species of paintbrush. This is the common spring paintbrush of desert and shrub steppe plant communities; other species are found in mountain meadows.

*Special Features*: Once established, Indian paintbrush will come back in the same spot year after year, with the first cheerful, red flowers of early spring, as well as the first reliable nectar for migrating hummingbirds.

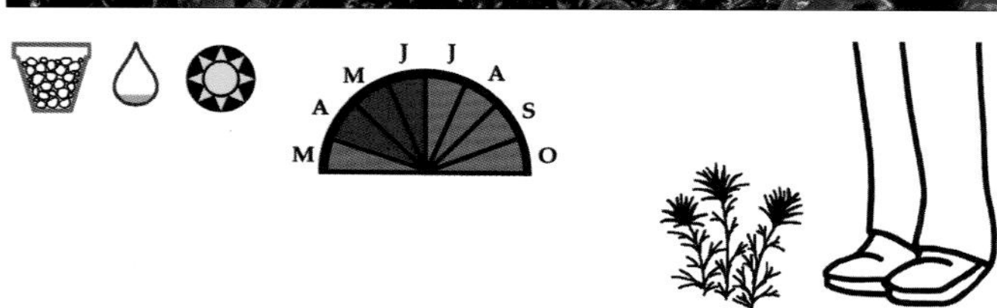

## Silver Buckwheat

***(Eriogonum ovalifolium)***

This eye-catching little plant features tight mounds of fuzzy, silvery green leaves, topped with flowering stalks that look like lollipops. The blossom puffs vary in color from cream or white through dark rose or even butter yellow. These turn rusty red as the seeds ripen. Silver buckwheat is easy to grow and tolerant of a range of soil types, though it will live longer in a lean, well-drained soil. It is an excellent rock garden plant, and it also works well as a perennial border plant or even as a drought-hardy ground cover. The plant keeps its shape and color and looks good even when not in flower. It volunteers readily from seed. If this is not desirable, just deadhead once the puffs turn rusty.

*Special Features*: Silver buckwheat is a classic example of a cushion plant, which is essentially a little shrub that is condensed into a tight shape. Cushion plants can survive in tough environments, including alpine tundra as well as deserts.

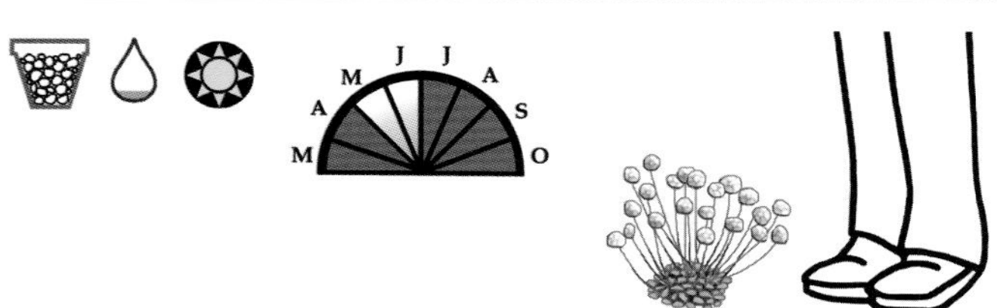

## Sundancer Daisy

***(Tetraneuris acaulis)***

This elegant, golden daisy features slender stems with single flowering heads that seem to dance above the mounds of bright green, grass-like leaves. One of the better-behaved members of its family, sundancer daisy is not much given to volunteering from seed, making it a good choice for more formal plantings. It looks magnificent in mass plantings and also combines well with plants like silver daisy or purple crazypea for a pleasing, polychrome effect. It is not fussy in its requirements, making it a good plant for people just getting started with native plant gardening. It occurs over a wide range of plant communities in nature, from desert and foothill rock gardens to alpine tundra, but always in the bright sunlight of open spaces.

*Special Features*: This plant shows an astonishing variability in flowering stalk height. Nine-inch stalks are the norm, but some tundra and badland races are less than an inch tall.

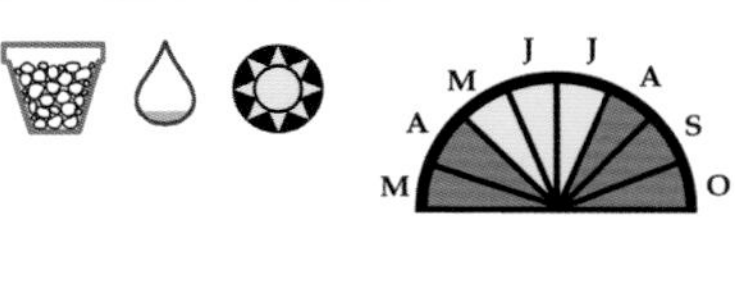

## Purple Crazypea

***(Oxytropis lambertii)***

This plant has a lot in common with Utah sweetvetch, including brilliant magenta pea blossoms, but it is a true desert plant that can succeed in minimal water landscapes. It can also tolerate medium water environments as long as the drainage is good. Its pale green leaves are held nearly upright, giving the plant a compact, tidy appearance. It puts on a show in late spring that is truly outstanding, and the straw-colored seed pods that follow are also subtly attractive. Purple crazypea rarely volunteers from seed. It is a well-behaved plant that looks good interplanted with Indian ricegrass, gooseberryleaf globemallow, and prince's plume.

*Special Features*: This plant is popular with big native bees when in bloom, though they often nearly weigh down the flowering stalks in their efforts to trip the entrance into the flower.

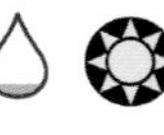

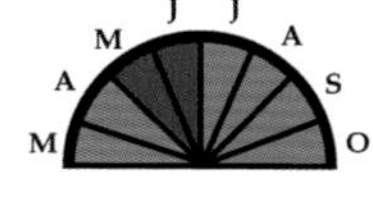

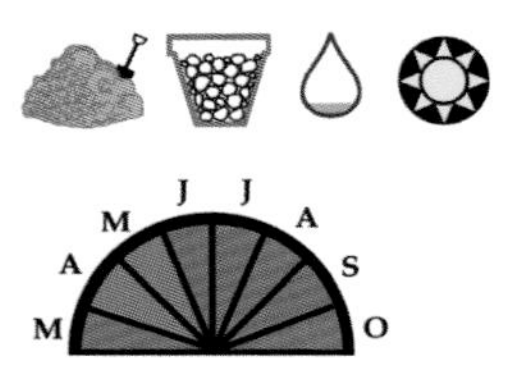

## Utah Penstemon

***(Penstemon utahensis)***

The combination of neon pink flowers and waxy blue foliage makes this plant a showstopper when in bloom. It is one of the earliest-flowering penstemons, making it especially welcome in the spring garden. Found in the driest, rockiest places in the southeastern part of our region, it is not tolerant of coddling. In fact, it can prosper in pure sand, a planting medium some experts recommend for many desert penstemons. Make sure the soil has excellent drainage and minimal organic matter, and give the plants plenty of room. A mature plant can have up to thirty flowering stalks, a sight that is unforgettable. Dwarf goldenbush and silver buckwheat make good companion plants.

*Special Features*: This plant is one of the suite of native species that are badland specialists. Badlands have heavy clay soils, but are located in such dry places that the excess water-holding ability of the clay is not a problem.

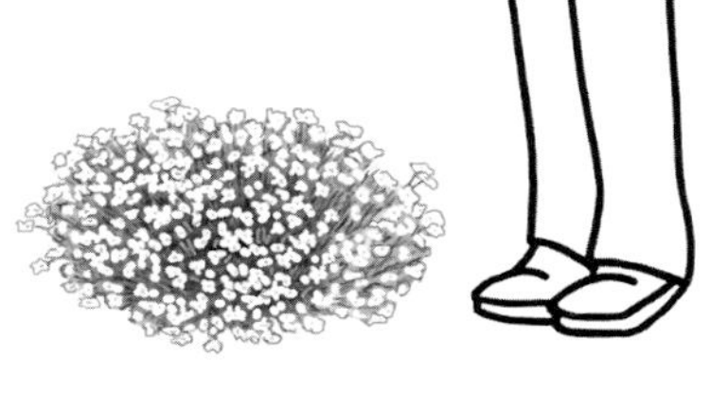

## Showy Sandwort

***(Arenaria macradenia)***

This handsome plant of rocky outcrops is a relative of domestic baby's breath, and the family resemblance can be seen in the white flowers. It has very fine, needlelike foliage that makes it rather inconspicuous when not in flower, though it does provide a sparse green backdrop for other plants. But the airy domes of flowers are the main attraction, lasting for several weeks in early to midsummer. This plant looks especially good flowering with sulfurflower buckwheat, and it can be used effectively in dry meadow plantings. It rarely volunteers from seed, so it can also be used in more formal settings. It is a long-lived, almost shrubby perennial that will bloom for many years with virtually no care.

*Special Features*: In nature, this plant occurs over a wide range of elevations, from low desert to alpine fell fields, but the common denominator is always rock. Fortunately, it is not nearly as picky as this preference for rock seems to imply.

## Indian Ricegrass

***(Achnatherum hymenoides)***

This distinctive, cool-season bunchgrass is one that many people can recognize, with its open, airy flowering stalks, threadlike green leaves, and seeds that look like little black BBs. It can thrive in hot, dry places, but it is equally at home on infertile soils in the foothills, making it an attractive addition to many desert, semi-desert, and foothill plantings. It looks best when grown in a lean, fast-draining soil and when given plenty of room to express its fountainlike growth form. Indian ricegrass is usually relatively short lived, especially when life is too good, but it is a prolific seeder and volunteers readily. It is best used in larger-scale, informal plantings, where it can replace itself from seed.

*Special Features*: Indian ricegrass seeds are edible and even tasty, and were a staple crop for native people of the region. Out in the desert, they are also the favorite food of kangaroo rats, who collect them by the thousands for later consumption, and plant many of them as part of the bargain.

## Gooseberryleaf Globemallow

***(Sphaeralcea grossulariifolia)***

Like many globemallows, this plant features orange flowers that resemble miniature hollyhocks, in this case borne along vertical stems above a mass of deeply lobed green leaves. It is broadly adapted but does best in lean, dry soils, where its flowering display can be quite showy. Where life is too cushy, it tends to grow mostly leaves. Globemallows produce abundant, long-lived seeds that generate a steady supply of volunteer seedlings. Clipping the stalks while the seeds are still green is a good way to prevent self-seeding, and it also can trigger another round of flowering if soil moisture is sufficient. Plant gooseberryleaf globemallow with other species of globemallow at your own risk—as mentioned before, globemallows hybridize freely.

*Special Features*: Many people never notice the delicate, sweet scent of globemallow flowers, which resembles a mix of orange blossom and cotton candy.

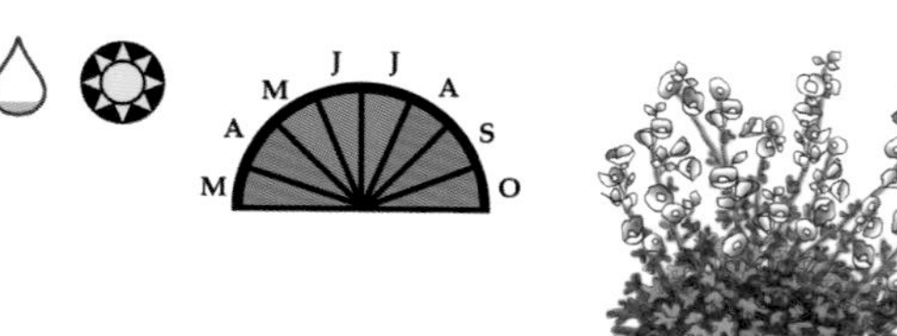


## Perennials: Minimal Water Use

**(Ranked Short to Tall)**

## Desert Needlegrass

*(Achnatherum speciosum)*

In nature, this elegant bunchgrass is usually found growing among the rocks in desert and semi-desert communities, and rocks do show it off to good advantage. But it is broadly adapted and tolerant of a range of soil types, and extends up into the foothills. It features vertical wands of feathery, platinum-colored fruits, which are even more luminous when backlit. Desert needlegrass needs some room to express itself, because the clumps increase substantially in size as the plant matures. It can provide good structure for a desert planting, as it keeps its flowering stalks for many weeks and stays green year round. It does have a tendency to self-seed, which makes it good for larger-scale, informal plantings.

*Special Features*: The individual fruits of this grass are like little works of art, so be sure and look at them closely. Be careful when handling them, though—the points are sharp.

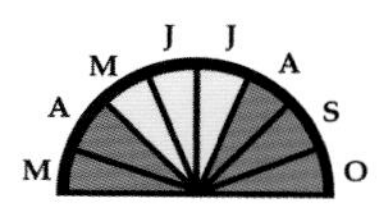

## Prince's Plume

*(Stanleya pinnata)*

This handsome, robust perennial is one of the hallmark plants of the desert and is especially characteristic of badlands communities throughout the Intermountain West and western Great Plains. It is a tough plant that is tolerant of clay, salt, and drought, but it usually does not live long in ordinary topsoil, especially if overwatered. Because of its large size at maturity, it is best used as a specimen (accent) plant. It does not volunteer much from seed, and can be used successfully in more formal settings. Its tall, golden flower spikes and blue-green foliage look especially fine when it is planted with alkali sacaton grass, another statuesque, salt-tolerant perennial.

*Special Features*: Prince's plume has large, nectar-rich flowers that attract an astonishing assortment of pollinators, so if you want to see unusual insects, keep an eye on this plant in blossom.

## Palmer Penstemon

***(Penstemon palmeri)***

Perhaps the tallest of our native penstemons, and certainly one of the most magnificent, Palmer penstemon combines large, waxy green, clasping leaves with long spikes of chubby pink flowers. The flowers, which bear a family resemblance to snapdragon flowers, have a fuzzy golden beardtongue as well as a memorable, sweet fragrance. Palmer penstemon is a true desert penstemon that does best in dry, lean, well-drained soils. In the right spot, a plant can live for many years. This species is best used as a specimen plant or mass planted as a screen. To prevent self-seeding, just clip the stalks after flowering. The foliage stays green all winter, except in the snowiest places.

*Special Features*: Palmer penstemon is the perfect bumblebee flower, and watching these mighty bees climb in for a sip of nectar is one of the most amusing aspects of a Palmer penstemon planting. Be sure to site the planting close to a seating area to enjoy the fragrance.

## Utah Ladyfinger Milkvetch

***(Astragalus utahensis)***

One of the earliest-blooming wildflowers throughout the Great Basin, this is just one of over a hundred milkvetches native to our region. It is a low, mat-forming plant that can be used as a drought-hardy groundcover or in a rock garden setting. In nature it is often found on road shoulders and in abandoned gravel quarries, habitats that indicate its preference for well-drained soils low in organic matter. It features woolly, mint green foliage with compound leaves typical of the pea family, and it is graced in spring with masses of large, pink to magenta blossoms. It does volunteer from seed, but usually not in great numbers. Cushion globemallow and fragrant evening primrose are good early-flowering companions for this plant.

*Special Features*: After flowering, the fuzzy seed pods look like a flock of little white chicks surrounding the mother plant. Be careful handling them—they have beaks that bite.

## Dwarf Goldenbush

***(Stenotis acaulis)***

This tidy little plant is found throughout our region, from semi-desert communities on up into the mountains, usually on sunny sites with shallow, rocky soils. In flower it forms a tight dome of yellow daisies that is very attractive to both people and butterflies. In fruit it forms fuzzy balls of seeds that do not disperse far, and it does not volunteer much from seed. It is easy to grow and tolerant of a range of soil types, though it does like good drainage. Its compact growth form suggests use in a rock garden setting or as a perennial border with other low-growing plants, such as silver buckwheat and cushion globemallow.

*Special Features*: Dwarf goldenbush is a plant for all seasons. Its bright green clusters of sword-shaped basal leaves keep it looking good even when it is not in flower.

## Shortstem Buckwheat

***(Eriogonum brevicaule)***

Shortstem buckwheat is one of the hidden treasures of the intermountain area. Its bright yellow flowers keep coming all summer, one of the longest bloom times of any native wildflower. Typically found in badlands communities in nature, it thrives in dry, exposed spots, and can tolerate salt and heavy clay soil. But it is a broadly adapted plant that performs equally well in richer, more moisture-retentive soils. The velvety, blue-gray foliage keeps it looking cool even on the hottest days. It is an excellent plant for rock gardens or for the front of a border. It combines well with other buckwheats and various low-growing perennials and carries the show started by these mostly early bloomers well into late summer.

*Special Features*: Shortstem buckwheat and other buckwheats are good flowers for dried arrangements—just clip and dry them when they are in full bloom. The bright color will last for years.

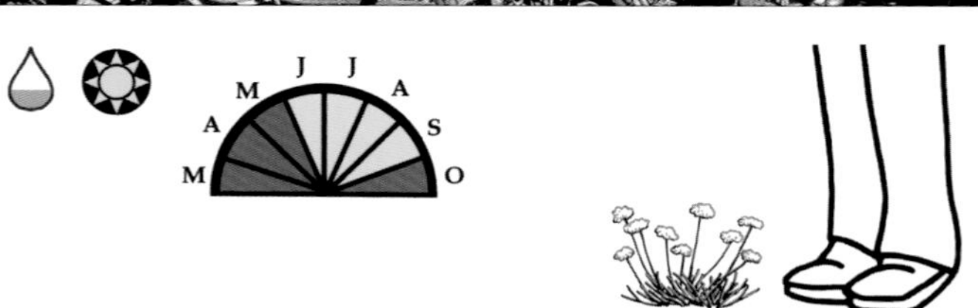


## Sulfurflower Buckwheat

***(Eriogonum umbellatum)***

The low, mounding form of sulfurflower buckwheat, combined with its shiny, dark green leaves, makes it an ideal species for rock gardens and more formal plantings. It can also be used in a prairie planting, though it needs plenty of light and space. It is not particular about soils, growing equally well in coarse, well-drained soils and those that are rich and moist. In flower, the plant forms a loose dome of bright sulfur yellow blossoms, sometimes tinged with red. This red becomes more pronounced in fruit, as the flower clusters turn rusty. Sulfurflower buckwheat does not volunteer freely from seed, though occasional new plants may be seen.

*Special Features*: Sulfurflower buckwheat is an evergreen plant that looks good throughout the year. The leaves frequently turn bright red in winter, adding extra color and interest to an often bleak season.

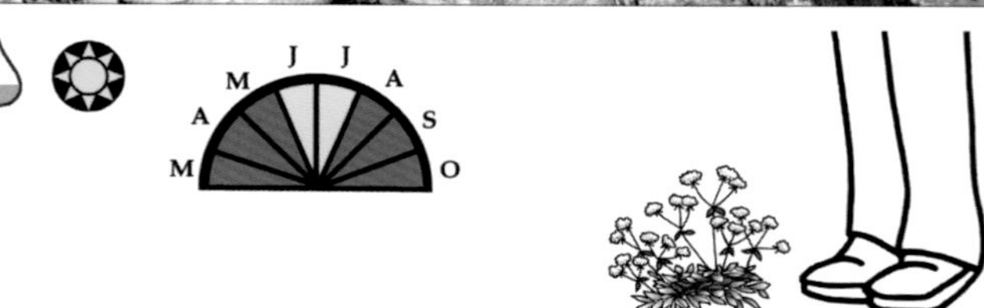

## Silver Daisy

*(Erigeron argentatus)*

Just one of many attractive daisy species, this characteristic southern Great Basin plant features lavender flowering heads, each on a slender stalk held above a basal tuft of silvery, strap-shaped leaves. It is easy to grow and not picky about soils, and it can be used in the medium water zone if the drainage is adequate. It is small enough for rock garden use, but can also be mixed to good effect with other wildflowers and grasses of the sagebrush steppe in an informal dry meadow setting. It looks especially attractive with sundancer daisy, which is about the same size but has contrasting bright green foliage and golden flowering heads.

*Special Features*: Silver daisy is a magnet for butterflies in the garden, so if you want to see a diversity of these attractive pollinators, try a silver daisy planting.

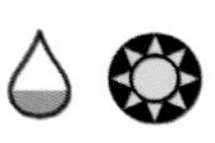

## Blue Grama

*(Bouteloua gracilis)*

Blue grama is a summer-active grass that is a dominant species on the Great Plains. In the intermountain area, it is confined mainly to the south, where summer rains are more reliable. Blue grama is a versatile grass, occurring naturally from the desert shrubland up into the ponderosa pine parkland. It may be a bunchgrass or a weak sodformer, depending on its origins. The bunchgrass form makes a wonderful specimen plant—the attractive flowering spikes persist into late fall. Blue grama is also useful as a substitute for cool-season turf grasses. It requires about a quarter as much water as Kentucky bluegrass, and can tolerate mowing and moderate foot traffic. It may also be left unmowed, and performs well in prairie mixes with spring and summer wildflowers.

*Special Features*: Blue grama has intriguing one-sided flowering spikes that have given it common names like eyelash grass and navajita (little razor).

## Shining Muttongrass

*(Poa fendleriana)*

Shining muttongrass is a cool-season bunchgrass that can add color and texture in a prairie, sagebrush steppe, or mountain brush setting, as well as in more formal plantings. It begins growth very early and stays green into the fall with little supplemental water. It has a compact growth form similar to blue fescue, but shining muttongrass has the advantage of being tolerant of both full sun and partial shade. Few native grasses are as shade tolerant as this one. Foliage color varies from bright green to pale blue, and the potential exists for selection based on this variation. Shining muttongrass is broadly adapted, not picky about soil, and very easy to grow. It is not a prolific seed producer, and behaves very well in a landscape setting.

*Special Features*: Shining muttongrass is named for its beautiful pearly pink to silver-green flowering heads, which appear very early in the spring.

## Utah Sweetvetch

*(Hedysarum boreale)*

In spite of its name, Utah sweetvetch is widely distributed in the Intermountain West, mostly in the sagebrush steppe and mountain brush communities. Especially drought-hardy forms can occasionally be found in desert communities, but specimens encountered in the nursery are from higher elevations and not likely to be quite so tough. Utah sweetvetch combines lush, green foliage with a spectacular display of magenta flowers in early summer. It makes an excellent understory plant for mountain brush or sagebrush steppe plantings, and also holds its own with grasses in prairie plantings. It shows only a moderate tendency to self-sow, and can thus be used in more formal perennial beds as well.

*Special Features*: Many peas look somewhat alike in flower, but Utah sweetvetch has very distinctive seed pods. They look like chains of flattened disks, each one containing a single seed.

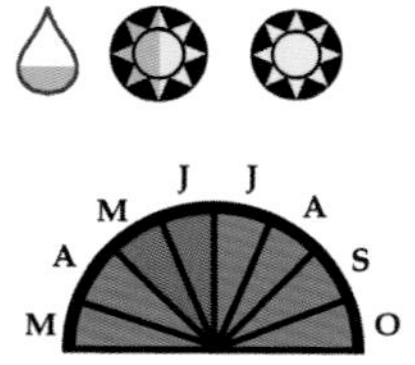

## Desert Four O'Clock

***(Mirabilis multiflora)***

This plant ranks among the legendary desert wildflowers. Its large green leaves and gargantuan growth form make it look almost tropical, but it is one of the toughest and most drought-hardy of native perennials. It blooms all summer long with little or no extra water. When in full flower in late afternoon, it is a glorious sight. Desert four o'clock is broadly adapted, but if life is too good, it will make more leaves than flowers. It does best in a lean, coarse soil. It should be planted as a specimen plant or in a bank with plenty of room between plants. The plants can live many years, and they will get bigger each year.

*Special Features*: The secret to drought hardiness for this plant is a large underground water-storage organ. It starts anew each year from this massive root when the soil warms up in late spring—it will begin growing later than you expect.

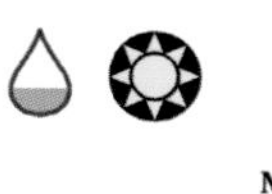

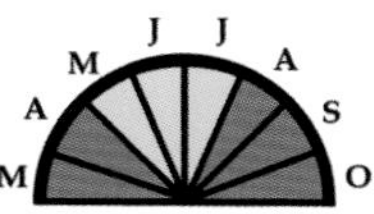

## Hopi Blanketflower

***(Gaillardia pinnatifida)***

Hopi blanketflower is native to the southeastern part of our region, the Colorado Plateau country. It features large heads with deeply notched, dark yellow rays and domed red centers, held above a sparsely leafy plant with deep green lobed leaves. It often grows in spare, sandy soils, but it is tolerant of rich soils, as long as they are well drained. It is best used in prairie plantings or as a shrubland or woodland understory species. Like all blanketflowers and most of their relatives, it has a strong tendency to seed itself, so it is best used in informal plantings.

*Special Features*: The seeds of Hopi blanketflower are popular with birds. They were also used to make a sort of peanut butter by indigenous people, who also used the plant as a medicinal.

## Lewis Flax

***(Linum lewisii)***

Lewis flax is found from the salt desert to the mountain meadows, but it is most characteristic of shrub steppe and foothill communities. It is one of the easiest native wildflowers to grow. The plants are sparsely leafy with fine, blue-green leaves, but when the flowers open, the plants become very noticeable. Flower color varies from sky blue to almost silver-white, and flowers are produced in great abundance over an extended period. Lewis flax is a prolific seeder and volunteers readily. It is a wonderful addition to informal prairie, shrub steppe, and mountain meadow plantings, where it combines well with Utah sweetvetch, showy sandwort, and Hopi blanketflower.

*Special Features*: Each Lewis flax flower lasts a only single day, and by the end of the season the ground beneath the plant is often littered with blue confetti.

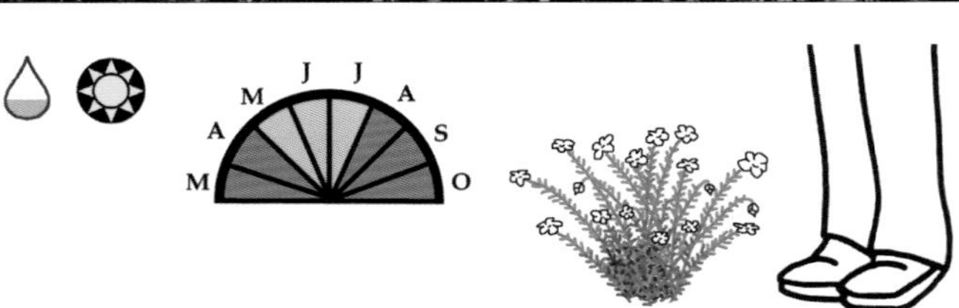


## Wasatch Penstemon

***(Penstemon cyananthus)***

Just one of many fine blue penstemons found in the Intermountain West, Wasatch penstemon is a plant primarily of the sagebrush steppe and mountain brush communities of the central part of the region, though it is also found in aspen understory and mountain meadow communities. It is somewhat more shade tolerant than most penstemons, as well as more tolerant of a variety of soil types. It can also cope better with extra water. Wasatch penstemon features bright green leaves and intensely blue flowers that are smaller than some other species but tightly clustered on the stem, for a very showy effect. It combines well with Indian paintbrush, sundancer daisy, and Utah sweetvetch.

*Special Features*: Wasatch penstemon tolerates competition from other plants better than most penstemons, and it can be used successfully in prairie and meadow plantings.

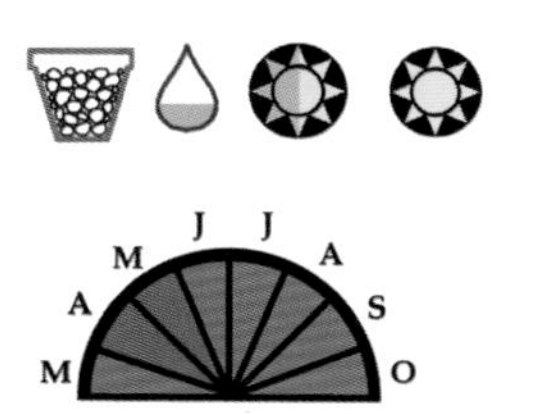


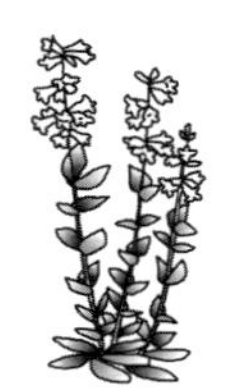

## Bluebunch Wheatgrass

***(Pseudoroegneria spicata)***

This grass, along with big sagebrush, forms the backbone of the sagebrush steppe community, and it is also a dominant species in the palouse prairie. No steppe or prairie planting would be complete without it. Fortunately, it is also a very beautiful plant, and one that is easy to grow. It is tolerant of a wide range of soil types and can live with both benign neglect and some overwatering. It provides texture, color, and structure, especially in mid- to late summer, when most of the other plants in these communities have already finished flowering. It is a prolific self-seeder, so it may occasionally be necessary to weed out seedlings to keep this species from overdominating a planting, especially on a fertile, well-watered site.

*Special Features*: The slender flowering stalks of bluebunch wheatgrass are a lovely sight when backlit in the late afternoon, and the plants manage to look good even during those January thaws, when all the other plants just look flattened.

## Bridges Penstemon

***(Penstemon rostriflorus)***

Bridges penstemon is a relatively common plant in semi-desert and foothill communities across the southern half of our region, from the Sierras to the Rockies. It can be a somewhat rangy plant, especially in overly fertile soils, but its elegant blossoms more than make up for its rather open structure. It has a broader soil tolerance than many penstemons, but it does look better when grown in a coarse, lean soil. It can be used in perennial beds or as part of the understory for semi-desert shrub and foothill woodland plantings. It looks especially good planted with other midsummer flowering species, such as littlecup penstemon, gooseberryleaf globemallow, and Lewis flax.

*Special Features*: Bridges penstemon is usually the last red-flowered penstemon to bloom in our area, making it a good companion plant for firecracker penstemon, another red-flowered species that blooms earlier in the summer, in a hummingbird garden.

## Littlecup Penstemon
***(Penstemon sepalulus)***

Littlecup penstemon is an upright, bushy plant with narrow, sea green leaves and long stalks of slender, deep lavender, snapdragonlike flowers. In the wild, littlecup penstemon is a specialist on steep, eroding banks in the sagebrush steppe and mountain brush zones, but it has turned out to be a very tractable garden plant, thriving in a variety of soils and experiencing very few problems. Even though it appears somewhat shrubby, it is best to cut this plant back close to the ground in the late fall, as flowering takes place from new shoots each year. Littlecup penstemon is best used in perennial beds and tall borders. It looks especially good with Bridges penstemon against a wall or rock backdrop.

*Special Features*: Littlecup penstemon is an example of an endemic plant, that is, a plant found growing wild in a very restricted area, namely the southern Wasatch Mountains in northern Utah. Fortunately, it performs well in gardens over a much wider area.

## Firecracker Penstemon
***(Penstemon eatonii)***

Firecracker penstemon is the common red-flowered penstemon throughout the southern half of the Intermountain region, and it has naturalized from roadside seedings in the north as well. It is a handsome plant, with large, shiny, dark green leaves and tall stalks of flared tubular flowers that hang down along the stalk. Firecracker penstemon can be a little tricky in the garden—it needs a well-drained soil low in organic matter. The good life makes for a short life for this plant. It is best used as a specimen plant or in a screen or mass planting with other tall penstemons. It will volunteer freely from seed—to prevent this, just cut the stalks after flowering is finished.

*Special Features*: Like most red-flowered natives in our area, firecracker penstemon attracts hummingbirds and depends on them for pollination. Watching the hummingbirds dueling for control of a firecracker penstemon patch in flower is great spectator sport on a summer afternoon.

## Alkali Sacaton Grass

***(Sporobolus airoides)***

This large bunchgrass is notable for its tolerance to salt, heavy soils, and subsurface moisture, but it is also quite drought tolerant. It can be used in minimal water landscapes if even a modest effort is made to provide it with harvested water—it is often seen growing along road shoulders, where it gets some runoff, in very dry places. Alkali sacaton is a beautiful grass that could serve as a handsome substitute for exotic ornamental grasses that tend to become invasive. A closely related plant, giant sacaton, is even larger and would be a suitable substitute for pampas grass. Alkali sacaton grass is best used as a specimen plant or in large massed plantings. It combines well with lacy buckwheatbrush, a shrub that is also tolerant of salt and clay soils.

*Special Features*: Alkali sacaton grass is especially beautiful in flower, when the finely divided, spangly flowering heads take on a luminous pink tinge.

## Mat Penstemon

***(Penstemon caespitosus)***

This little penstemon is quite different from its taller cousins. It is a mat-former that pins itself to the ground by rooting at the nodes as it spreads. Its small, dark green, almond-shaped leaves stay green all summer. It makes a beautiful ground-cover plant, especially when used to fill the spaces between the flagstones of a pathway, and it can take some light foot traffic. An added bonus is the mid-spring display of miniature blue, snapdragonlike flowers with bright orange "beardtongues" sticking out of their little faces. This plant is not picky about soil, but it does benefit from occasional watering during the heat of the summer.

*Special Features*: Tiny flowers like those of mat penstemon are pollinated by equally tiny native bees, and some of these, such as the metallic turquoise *Osmia* bees, are as pretty as the flowers.

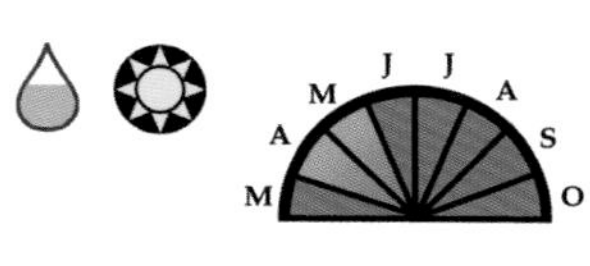

## Rosy Pussytoes

***(Antennaria rosea)***

This plant often forms patches at the edge of the oaks and maples in the foothills, and it is also found in open meadows higher in the mountains. It is a good choice for a rock garden or as a ground cover. It can tolerate light shade, which makes it useful as an understory plant in mountain brush plantings. When not in flower, it is very short, only an inch or so tall, but the flower stalks rise above the cottony mat of basal leaf rosettes. The height of the stalks is variable, ranging from a couple of inches to nearly a foot. A ground-cover planting of taller forms of rosy pussytoes can easily be tidied up with a string trimmer once flowering is finished, and the planting can take light traffic when not in flower. This plant likes a well-drained soil, but it appreciates a little extra water during hot weather.

*Special Features*: Rosy pussytoes is almost cuddly to the touch, and is named for the soft clusters of flower heads that do indeed feel a bit like the toes of a pussy cat.

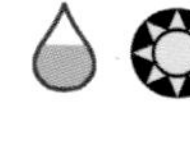

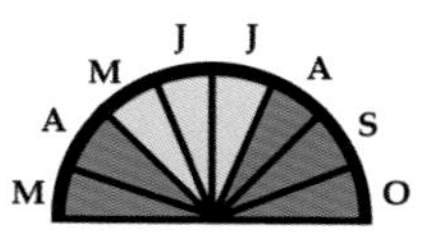

## Trailing Daisy

***(Erigeron flagellaris)***

Trailing daisy gets its name from its habit of producing runners, which then root at the tips like strawberry runners. It can use these runners to form large patches, making it a useful ground-cover plant in the foothill and mountain water zones. It does well in either full sun or partial shade, but will need less water if not exposed to afternoon sun in the summer. It appreciates rich, water-retentive soil. The flowering stalks vary in height, but taller forms in ground-cover plantings can be trimmed after flowering with a string trimmer.

*Special Features*: It is quite possible to manage a trailing daisy planting as a substitute for turf, and the planting will require far less water than lawn, as well as looking very pretty when in flower. The rest of the time it forms a low green mat that can take some light foot traffic.

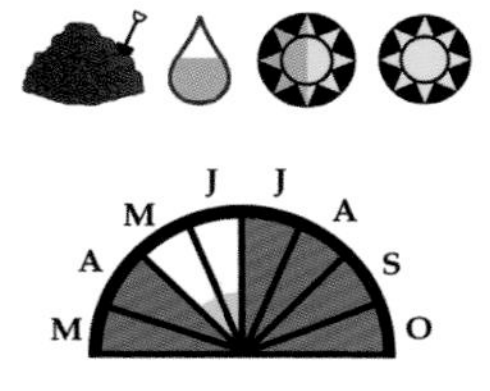

## Leo Penstemon

***(Penstemon leonardii)***

Leo penstemon is a little-known species that has a relatively small distribution, centered in northern Utah, but it is a broadly adapted plant that could be used successfully throughout our region. It features a low, mounding growth form quite different from that of most penstemons, deep green foliage, and a profusion of electric blue, snapdragonlike flowers in spring. It is the amazing blue color of the flowers that is its chief selling point, though the plant does provide masses of green foliage color through the growing season, especially with a little extra water. This is a great species for a perennial border, and it could be used as a low-maintenance ground cover as well. It would also be a good addition to a prairie planting. Unlike most penstemons, it does not volunteer readily from seed.

*Special Features*: Leo penstemon flowers at the same time as Indian paintbrush, and makes a good buddy plant. When they are planted together, the combination of bright blue and bright red flowers is spectacular.

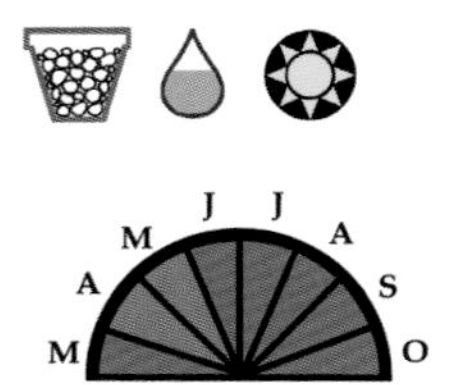

## Lavenderleaf Sundrops

*(Calylophus lavandulifolius)*

This low-growing plant has rather sparse, fine-textured foliage, but its deep yellow flowers are quite large, so that in bloom the plant seems to be mostly flowers. The blossoms start as satiny red-and-green striped buds, open to almost square, four-petaled flowers, and fade through shades of orange and red as they age. Lavenderleaf sundrops can keep up this parade of blossoms all summer long. A plant of rocky, open country, it needs full sunlight to prosper, but it is tolerant of a range of soil types as long as the drainage is good. It is a great rock garden plant. It can also be used in traditional perennial beds and as an understory species in open foothill woodland plantings.

*Special Features*: This plant is strongly perennial and will live many years under good conditions. It is not much given to self-seeding, so plant it where you need it to be.

## Flaxleaf Penstemon

*(Penstemon linarioides)*

The leaves of flaxleaf penstemon are narrow and strap-like, similar to those of Lewis flax. It has a low, mounded form that works well in a perennial border or rock garden setting, though it does have a strong tendency to self-seed. It also makes a good understory plant in the openings of mountain brush and foothill woodland plantings, and it could be used as an informal ground cover. Flaxleaf penstemon can hold its own against perennial grasses, in spite of its small stature. It combines well with lavenderleaf sundrops, Bridges penstemon, and shortstem buckwheat. Flower color varies in this species from deep sky blue to almost white, but the flowers are usually a soft baby blue.

*Special Features*: Flaxleaf penstemon is one of the last penstemons to flower each summer, providing a welcome splash of color in midsummer landscapes. Its pale blue flowers are very pretty against the fine screen of dark green leaves.

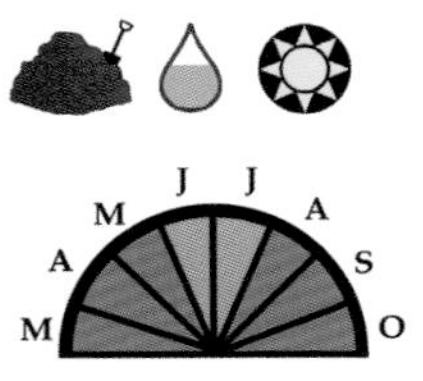

## Butterfly Milkweed

***(Asclepias tuberosa)***

It may be a surprise to learn that this familiar garden plant is an intermountain native. It grows naturally throughout the eastern and central United States, and enters our area in the mountain brush and ponderosa pine communities of the Colorado Plateau country. Butterfly milkweed rarely seeds itself, making it a welcome plant in traditional gardens. It can also be used as an understory plant in pinyon and ponderosa pine openings, where it combines well with little bluestem and blue grama. It prefers a rich soil and regular water in hot weather. It will reward your attention with a magnificent display of bright yellow-orange flowers in mid- to late summer, at a time when little else is blooming.

*Special Features*: Butterfly milkweed is well named. It is a great plant for a butterfly garden, attracting a variety of different species. Later, you can enjoy the beautiful flight of the milkweed seeds themselves, without worrying much about weeding volunteers.

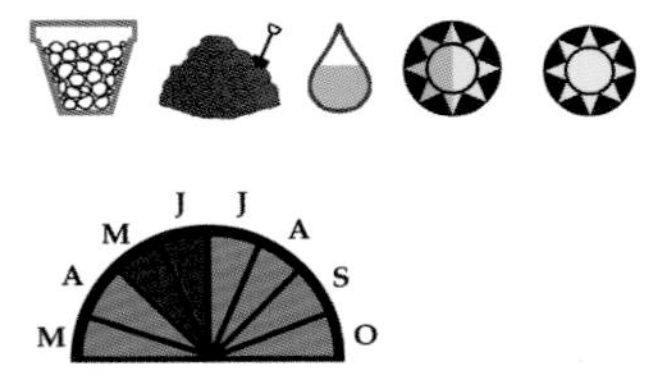

## Whipple Penstemon

***(Penstemon whippleanus)***

An unusual penstemon of rocky soils at the edges of mountain conifer forests, Whipple penstemon has bright green, finely toothed leaves and flowers that are nearly tubular and close to the color of grape Kool-aid. Occasional plants have dusty mauve or maroon flowers. They occur in clusters at intervals along the flowering stalk, for a decidedly showy effect. This penstemon is a delicate plant, quite different from its robust relatives of lower elevations, and it needs to be mass-planted or given a prominent spot to show off its rather exotic beauty, which passes all too quickly. It prefers light shade, a rich but not heavy soil, and a mulch of conifer needles.

*Special Features*: Whipple penstemon can be used as the centerpiece of a shade garden featuring firechalice and Rocky Mountain and western columbines as companion plants.

## Little Bluestem

***(Schizachyrium scoparium)***

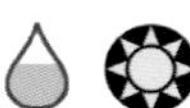

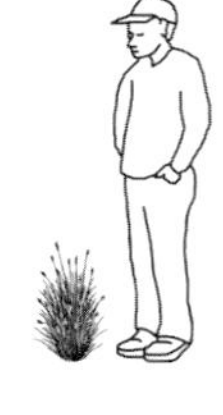

Little bluestem is best known as a tallgrass prairie plant, but it has a wide distribution in North America, entering our region in the Colorado Plateau country. It is a warm-season grass that flowers in late summer and holds its seed heads well into the fall. These feathery seed heads, combined with rich green foliage and a lovely, fountain-like growth form, make it one of our most beautiful native grass species. It works well as a specimen plant or in a bank, and it can also be used as an understory species in foothill woodland and ponderosa pine plantings. It needs summer rain to prosper, so it is not a suitable choice for a palouse prairie planting.

*Special Features*: The foliage of little bluestem provides wonderful fall and winter color, turning various shades of yellow, orange, and red. The cultivar 'Blaze' was selected for exceptional fall color, and does well in our region.

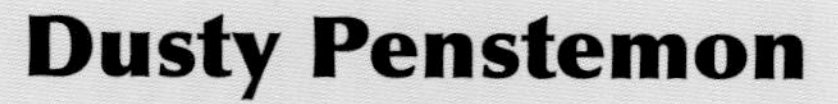

## Dusty Penstemon

***(Penstemon comarrhenus)***

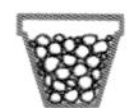

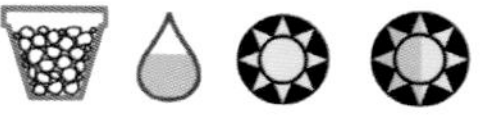

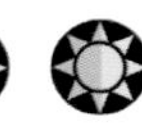

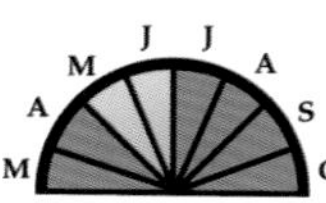

This elegant but little-known plant is native to the Colorado Plateau and southeastern Great Basin, where it is often the most common penstemon species at middle elevations. It features a tall, willowy growth form, narrow, pale green leaves, and long stalks of large, baby blue flowers with pale pink throats. Occasional plants have shell pink flowers. Dusty penstemon makes a beautiful specimen plant, and it would fit in well in a tall border planting. It can also be used in mountain meadow plantings or as an understory species in mountain brush and aspen parkland communities, as it can tolerate light shade. It combines well with Utah sweetvetch and mountain puccoon.

*Special Features*: Dusty penstemon is not as finicky about soil as most penstemons, and can handle more water than most, making it relatively easy to grow.

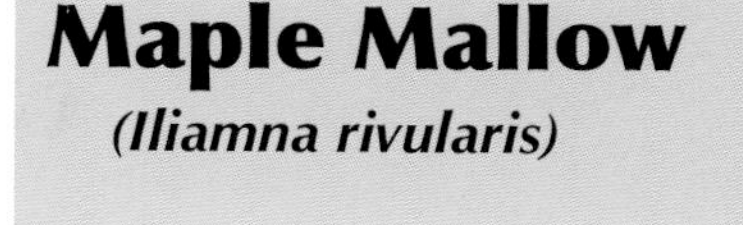

## Maple Mallow

***(Iliamna rivularis)***

This robust plant is one of the showiest wildflowers of middle elevations throughout the intermountain region. It combines large, maplelike leaves with spikes of pale pink to deep rose flowers that resemble miniature hollyhocks. It is best used as a specimen plant, as individual plants can get quite large, especially in favorable situations. It could also be used in a tall border, in a mass planting used for a screen, mixed with other tall flowers and grasses in a mountain meadow, or as part of a mountain brush or aspen understory planting. The plants do not get so large when grown in partial shade or on drier sites.

*Special Features*: You will rarely see more than a few maple mallow plants at a time in the wild, except after a fire, when large stands spring up from seeds that have stayed in the soil for decades, waiting for the heat of a fire to make them germinate.

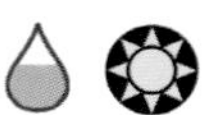

## Basin Wildrye

***(Leymus cinereus)***

The largest and perhaps the most spectacular of our native bunchgrasses, basin wildrye is widely distributed throughout the West. It is an excellent choice for a specimen plant or a tall screen planting. It can also be used as a component of a palouse prairie or sagebrush steppe community, though it will stay smaller on these somewhat drier sites. Basin wildrye keeps its structure year-round, and it is especially attractive in high summer when the foliage is a deep green that contrasts beautifully with the lime green flower spikes. Later the whole plant turns straw-colored. Given enough room, basin wildrye combines well with big sagebrush, oakleaf sumac, alderleaf mountain mahogany, and other large shrubs of the foothill zone.

*Special Features*: The "basin" in basin wildrye is short for the Great Basin, the region where this plant reaches perhaps its greatest abundance. It was so abundant there in pre-settlement times that it was harvested as a grain crop by the indigenous people.

## Prairie Smoke

***(Geum triflorum)***

This little wildflower is commonly encountered in mountain meadow communities, and it is right at home in the rich soil of a traditional garden bed. It can be used as a low border, or in an informal meadow planting with trailing daisy and flaxleaf penstemon as companions. Its leaves, found mostly at the base of the plant, are long and narrow, bright green, and deeply lobed along the edges. The rest of the plant, including the stems, the flowering stalks, and the flowers themselves, are an attractive dusty rose color. The flowers resemble small bells that nod charmingly from branched flowering stalks.

*Special Features*: Prairie smoke also goes by the names prairie duster and old man's whiskers. All these names refer to the feathery pink tufts of fruits, one of the most attractive features of the plant.

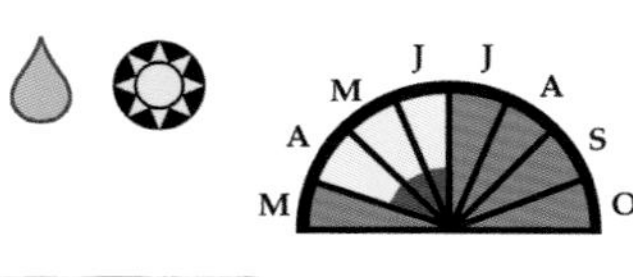

## Little Beebalm

***(Monardella odoratissima)***

The compact mounds of little beebalm can be found growing along protected canyon walls in the foothills, but this plant really comes into its own on open, rocky mountain slopes, where it can form extensive stands. In bloom, the plant is completely covered with lilac balls of flowers, putting on a beautiful display and attracting an interesting array of pollinators. Foliage color varies from bright green to pale gray-blue. Little beebalm is best used as a border plant or in the rock garden. It keeps its structure throughout the growing season and is only a modest self-seeder, making it suitable for more formal settings.

*Special Features*: If you brush your hand over the foliage, the sweet, menthol scent will tell you right away that this plant is a wild relative of mint.

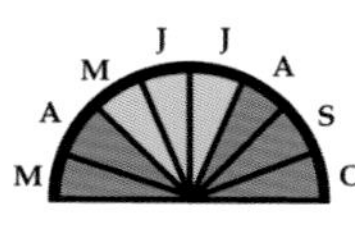

# Perennials: High Water Use
**(Ranked Short to Tall)**

## Mountain Puccoon

***(Lithospermum multiflorum)***

A little-known species from the southern mountains within our region, mountain puccoon is a wonderful addition to mountain meadow and aspen parkland communities, and it can also hold its own in more formal plantings. It features an upright growth form, narrow, bright green leaves, and sprays of golden yellow trumpet flowers that have a sweet fragrance. Mountain puccoon rarely self-seeds. It is not fussy about soil and can tolerate moderate shade. It combines well with dusty penstemon, showy daisy, and sticky geranium, plants that are often found growing together at the edge of the aspens on the high plateaus.

*Special Features*: Another name for the puccoon genus is stoneseed, and when you see the large, shiny white seeds of this plant, this name will make sense. The seeds look as if they are made of porcelain.

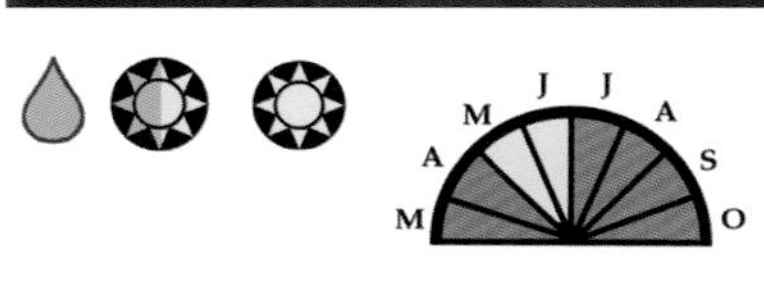

## Firechalice

***(Epilobium canum)***

Various forms of this plant are found throughout the southern part of our region, as well as to the south and west. Its usual haunts are shady canyon walls and rocky outcrops at middle to high elevations, but it has proven to be a remarkably versatile and broadly adapted garden plant. It spreads from underground runners to form patches, but the process is slow and not difficult to control. Firechalice can handle full sun or partial shade, and it will thrive in both rich, organic soils and spare, sandy ones. It flowers from late summer into the fall and can be used as an accent plant, in a bank or border planting, or as an informal ground cover. It does well planted as an understory between shrubs like squaw apple, mallowleaf ninebark, and mountain snowberry.

*Special Features*: Other common names for this plant are zauschneria and hummingbird trumpet. Its brilliant red-orange flowers provide the last nectar source of autumn for hummingbirds on their southward migration.

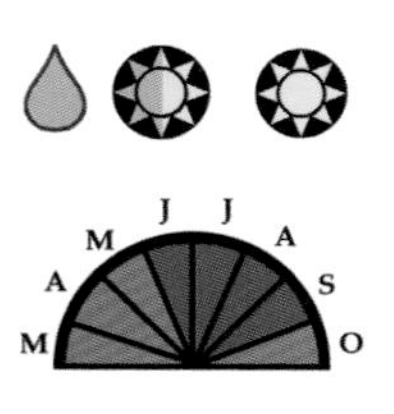

## Rocky Mountain Columbine

***(Aquilegia coerulea)***

One of the most beloved of all western wildflowers, this is also one of the easiest to grow. The large, white or two-tone white and pale blue blossoms are every bit as showy as any cultivated columbine species. Rocky Mountain columbine thrives in dappled shade, and it combines well with showy daisy, blooming sally, and mountain puccoon in plantings for aspen parkland or mountain meadow communities. It also performs well in more formal settings, though it will seed itself freely if the seedstalks are not removed when still green. The plants are also pretty when not in bloom, with an abundance of blue-green foliage that forms an attractive mound.

*Special Features*: Columbines are notorious for their tendency to hybridize, so if you plant more than one species, it is best not to let the plants self-seed, as the mixed-parentage offspring are rarely as pretty as the parents.

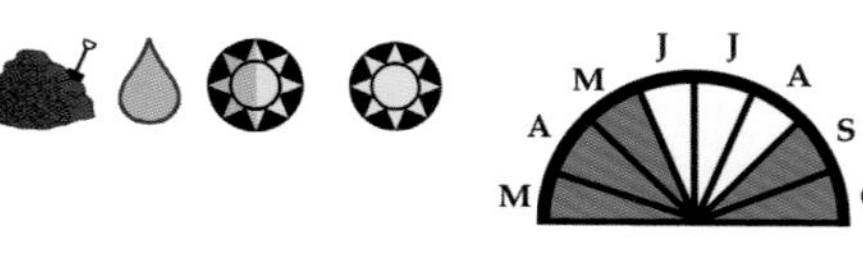

## Western Columbine

***(Aquilegia formosa)***

Western columbine is an elegant plant that does best in partial shade and rich, moist soils. In nature, it is almost always found at streamside or near springs, but it is not hard to grow in a garden. It has dainty, red-and-yellow, nodding blossoms on long stems over a compact basal mound of blue-green foliage. It performs well in informal as well as formal settings, and it can also be used in aspen parkland understory plantings. Western columbine will self-seed, though not as freely as domestic columbines. It hybridizes readily with other columbines, so do not plant it with other species if you want the self-seeded offspring to resemble their parents.

*Special Features*: Like all columbines, western columbine has petals with long spurs that hold the nectar reward for long-tongued pollinators like hummingbirds and hawk moths, while the stamens are thrust in a mass out of the front of the flower, where pollinators will be sure to contact them.

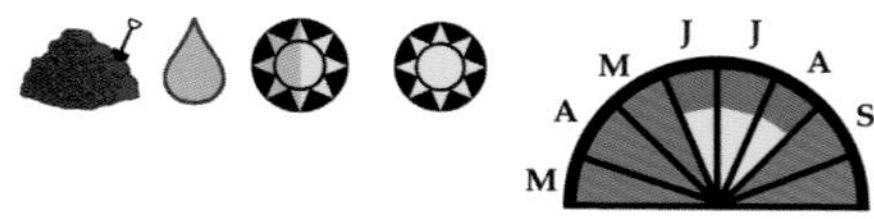

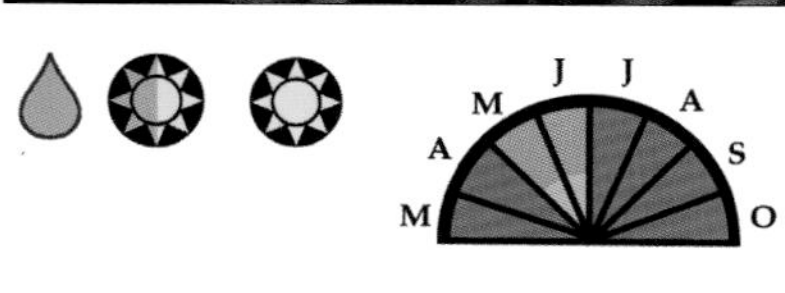

## Showy Daisy

***(Erigeron speciosus)***

Showy daisy is a characteristic species of aspen parkland and mountain meadow communities throughout our region. It is a robust plant that can put on an impressive flowering display, especially when grown as a specimen plant or in a bank in full sun. It also combines well with sticky geranium, maple mallow, and meadow fire in mountain brush, aspen understory, and meadow plantings. It features rounded masses of large daisy flowers with deep lavender rays and yellow centers. Showy daisy has bright green foliage that provides structure late in the season, after flowering is finished. It should be cut to the ground in fall. It spreads by short underground runners as well as by self-seeding, but it is not particularly assertive in this regard.

*Special Features*: The Europeans have developed several cultivars of this magnificent plant. This happens with many of our natives—no one here thinks they are anything special, but gardeners in other places value them highly.

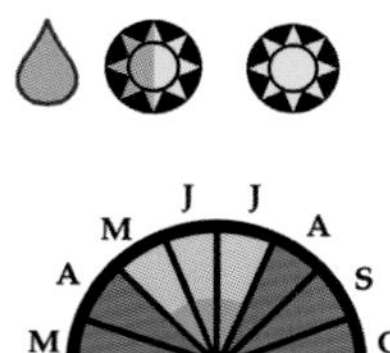

## Meadow Fire

***(Hymenoxys hoopesii)***

With its clusters of large, deep chrome yellow daisy heads, this plant is one of the showiest species of mountain meadow communities. The narrow, petal-like rays are bent downward, giving the heads a characteristic shaggy look. This plant does best in full sun but can also grow in dappled shade. It looks wonderful in massed plantings and as a companion plant for sticky geranium, tall larkspur, and blooming sally. The rather lush, bright green foliage is an added attraction. Meadow fire is not a prolific seeder in the garden and can be used in formal settings. It has been known in the past by some peculiar common names, including owl's claws and orange sneezeweed, a name that refers to its use as a substitute for snuff. We decided to christen it with a name worthy of its considerable beauty.

*Special Features*: Meadow fire is a great plant for the butterfly garden. Painted ladies seem to be especially attracted to its large, pollen-rich heads.

## Sticky Geranium

***(Geranium viscosissimum)***

This handsome plant is one of the easiest and most reliable natives for home landscapes. It features a basal clump of large, bright green, almost round leaves that are deeply lobed, and bright magenta flowers held above the leaves on branched flowering stalks. An added attraction is the brilliant red foliage color in the fall. Sticky geranium is not fussy about soils, though it does benefit from added organic matter. It thrives in full sun or partial shade. It is a modest self-seeder that can be used as a specimen plant, in perennial borders, or as a member of mountain brush, aspen parkland, and mountain meadow communities.

*Special Features*: The fruits of sticky geranium are fascinating from an engineering perspective. The seeds, held down by force at the base of the long central style, are catapulted from the plant when the dry style segments spring upward. Seedlings can show up in odd places.

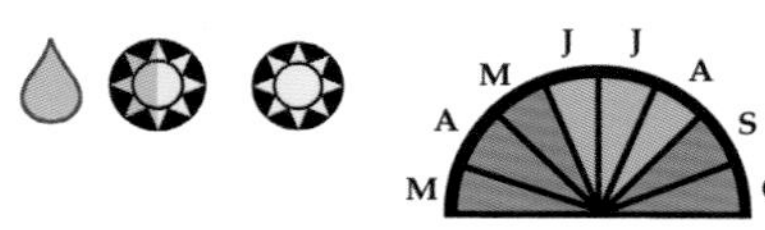

## Blooming Sally

***(Chamerion angustifolium)***

Very widely distributed in western and northern North America, blooming sally is a well-known wildflower, especially in the northern part of its range, where it is often very abundant. It is a slender plant with willowlike leaves, and its name comes from this resemblance to a blooming *Salix* (willow). It flowers in late summer, and its tall wands of deep magenta flowers look wonderful in massed plantings or with tall larkspur and meadow fire. Blooming sally spreads by underground runners and by self-sowing, but unless the site is very wet, the rate of spread will be slow and easily controlled, especially in partial shade.

*Special Features*: Fireweed is another name for this plant. It is a fire-follower, often occurring in large stands in the openings created by forest fires. Its tiny seeds are dispersed by the wind, enabling it to travel great distances to find new openings.

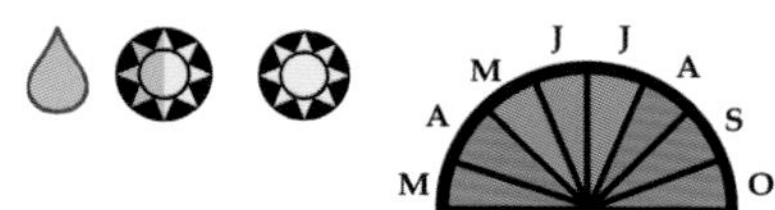

# Perennials: High Water Use

**(Ranked Short to Tall)**

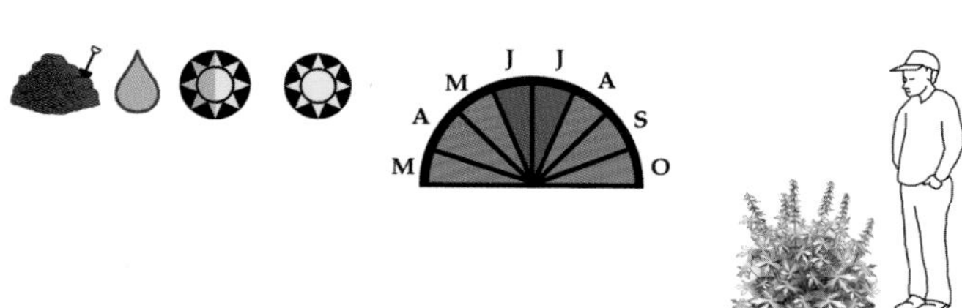


## Tall Larkspur

***(Delphinium barbeyi)***

This statuesque plant is similar to domestic delphiniums. It is perhaps not quite as showy, but it can thrive on much less water. It is an abundant species in aspen parkland and mountain meadow communities throughout our region. Tall larkspur features long spikes of deep blue-violet flowers over clumps of bright green basal leaves. Its large leaves are deeply lobed, almost succulent, and quite attractive even when the plant is not in flower. Tall larkspur combines well with other mountain meadow species like blooming sally, showy daisy, and meadow fire. It forms large clumps over time and makes a beautiful specimen plant or formal border display.

*Special Features*: Like its relatives the columbines, tall larkspur has flowers with nectar spurs to encourage pollinators. Native bumblebees are frequent visitors.

## Shadscale

***(Atriplex confertifolia)***

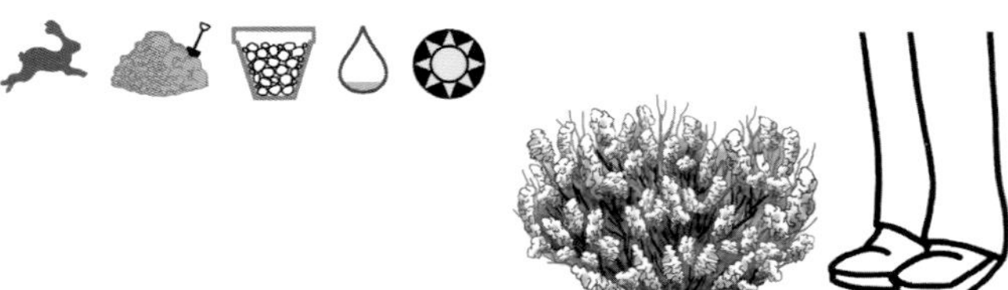

This little silver-green shrub is one of the toughest and best for minimal water use landscapes, tolerating salt, heat, and drought. It has small, nearly round leaves that have a soft shine and a pleasing, compact growth form. It grows quickly to mature size and often sets fruit the year after planting. Shadscale is definitely a plant that needs tough love—it hates organic matter and too much water. Do not try to use it in foothill or mountain plantings. Planted in the right place, it is low-maintenance and long-lived. Once established, it will not need supplemental water. It combines well with winterfat, desert sage, and green Mormon tea. Shadscale is quite thorny and hard to weed, so aim to plant it where weeds will not be an issue.

*Special Features*: The fruits of shadscale ripen in autumn and are one of the most attractive features of the plant, turning satiny pastel shades of pink, rose, and orange.

## Winterfat

***(Krascheninnikovia lanata)***

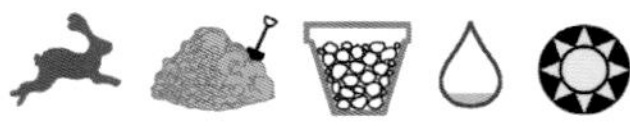

Winterfat is a fast-growing little shrub that thrives in minimal- and low-water-use landscapes. It has soft, white foliage, with small, upright leaves that look a bit like little rabbit ears. It is more tolerant of extra water and fertility than most desert shrubs, though it can get floppy if life is too good. The most appealing feature of the plant is the luminous white wands of cottony fruits that form in late summer and persist through the fall—these look especially beautiful in backlight. Winterfat volunteers freely from seed, especially on less dry sites. It looks good planted with lacy buckwheatbrush and shadscale for a handsome autumn display. Combine with prince's plume, Palmer penstemon, and desert sage for color through the season.

*Special Features*: Winterfat gets its odd name from the fact that it is a palatable and nutritious browse for sheep in the wintertime. Another name for the plant is whitesage.

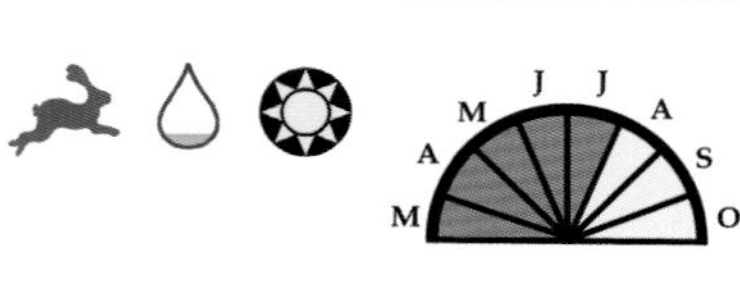

## Lacy Buckwheatbrush

***(Eriogonum corymbosum)***

This striking and unusual plant comes in a variety of shapes and sizes, but all feature soft, pale leaves and intricately branched flowering stalks. These give an airy but substantial, domelike structure to the plant. The plants flower in late summer to fall, and flowers vary from cream to dark rose pink or bright sulfur yellow. The flowers persist on the plants and turn rust colored, and eventually the whole dome turns rusty. This dome persists through winter and adds great structure and color to the winter garden. There is no need to clip these flowering stalks. In spring, the new growth will come right through the old. Lacy buckwheatbrush is very forgiving, especially considering its desert origins. It will thrive in rich soil or poor, with or without extra water. It just needs plenty of room to express itself.

*Special Features*: Lacy buckwheatbrush is the all-season plant *par excellence*. It looks great as a specimen plant, a hedge, a border, or a mass planting, and it combines especially well with desert sage.

## Desert Sage

***(Salvia dorrii)***

This hidden treasure of the desert looks a lot like many other little gray-green shrubs—except when it flowers. In blossom, it becomes literally covered with hundreds of spikes of royal blue flowers, each held inside a purple bract, for a very showy bicolor effect. The mass of flowers attracts many interesting native pollinators. The pale green leaves of desert sage are small and rounded, with a thick, almost leathery texture, and the plant is a true evergreen, providing structure and gray-greenery year round. A pungent scent somewhat like cooking sage is another nice feature. Desert sage needs a lean, well-drained soil to thrive, and will almost never need watering after establishment. It combines well with green Mormon tea, winterfat, and lacy buckwheatbrush.

*Special Features*: Desert sage makes a good host for Indian paintbrush. The contrast of deep red with royal blue and purple is spectacular. Add sundancer daisy to the mix for a stunning polychrome effect.

# Woody Plants: Minimal Water Use
**(Ranked Short to Tall)**

## Datil Yucca

***(Yucca baccata)***

Datil yucca is a handsome plant of the southwest plateaus, entering our region from the south, but it is fully cold-hardy. Its heavy, swordlike, blue-green leaves are borne in a massive clump. They feature beautiful curling white fibers along their edges and sharp, hard tips. The very large flowers of datil yucca are borne along short, stout stalks not much longer than the leaves. They hang downward in clusters of waxy, cream- and rose-colored bells. Datil yucca is a broadly adapted plant that is easy to grow, though it does prefer a well-drained soil. It makes a beautiful structural contrast planted with finer-textured shrubs like sand sagebrush, Apache plume, and rubber rabbitbrush.

*Special Features*: Datil is the Spanish word for "date," and refers to the fleshy, sweet fruits. These are rarely seen in cultivation because of the need for a specialized pollinator, the yucca pronuba moth. Another common name for this plant, banana yucca, also refers to these fruits.

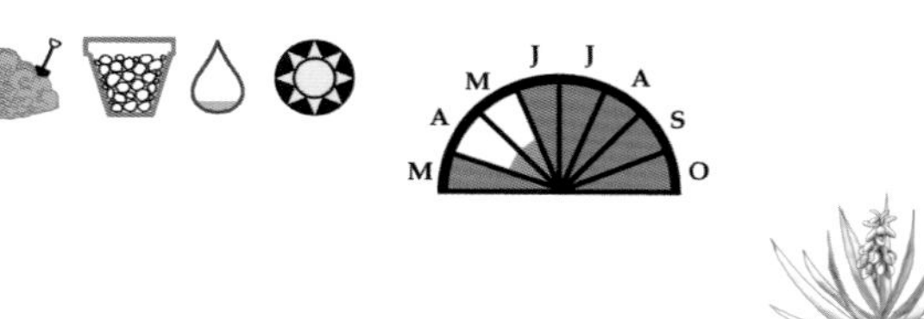

## Dwarf Yucca

***(Yucca harrimaniae)***

Dwarf yucca ranges further north in our region than any other yucca species. It forms compact clusters of narrow, sharp-tipped leaves with showy flowering stalks that are carried well above the leaf clusters. These clusters are sometimes very small, about baseball size, hence the name dwarf yucca. The plants more commonly have leaves up to a foot long. The waxy, cream-colored flowers are nearly round in outline and hang down along the stalks. These are followed in the wild by pods that are dry and that crack open to reveal flat black seeds stacked like coins in each chamber. Dwarf yucca looks good planted with green Mormon tea, Fremont barberry, and Indian ricegrass. It needs some room to express itself in order to look its best.

*Special Features*: Yucca plants can flower multiple times from the same leaf cluster, unlike their relatives, the century plants. Dwarf yucca also tends to propagate itself by forming "pups" at the base of the mother plant.

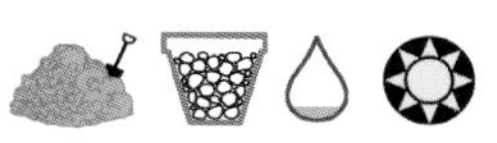

## Sand Sagebrush

***(Artemisia filifolia)***

This graceful and elegant plant has feathery, sea-green foliage on branches that tend to arch and droop downward as the plant grows older. The foliage color takes on an almost bluish hue against the red sands that are often its habitat in nature, and it also contrasts nicely with the shreddy, almost black bark of the trunks. This evergreen plant provides excellent structure and color throughout the year. Though confined in nature to sandy soils, it does not require sand to prosper. It does need a lean, well-drained soil and plenty of sunshine, however. This is not a plant for foothill and mountain precipitation zones, but it can form a magnificent backbone plant for a minimal-water landscape. It looks wonderful with datil and dwarf yucca, green Mormon tea, Palmer penstemon, Hopi blanketflower, and Indian ricegrass.

*Special Features*: Like all its relatives, sand sagebrush has strongly scented foliage, in this case a sweet, almost menthol-like fragrance that is especially noticeable after a rain.

## Big Sagebrush

*(Artemisia tridentata)*

Perhaps the best-loved and certainly the most maligned native shrub, big sagebrush is the signature plant for the entire region. Fortunately, this plant is easy to live with, both for people and for other plants. It provides a fine, pale green backdrop that sets off the colors and textures of smaller shrubs, grasses, and wildflowers. It is especially valuable as a source of winter structure and color. There are many forms, including basin big sagebrush, that can grow to ten feet or more. Wyoming big sagebrush is much smaller and is best for low-water-use landscapes, while mountain big sagebrush, also small and exceptionally sweet-scented, is best for foothill and mountain landscapes.

*Special Features*: We plant Indian paintbrush with big sagebrush, which is deep-rooted and will provide water to keep its companion flowering longer into the spring. Basin wildrye, bluebunch wheatgrass, Utah sweetvetch, and Lewis flax also make great companions for big sagebrush.

## Green Mormon Tea

*(Ephedra viridis)*

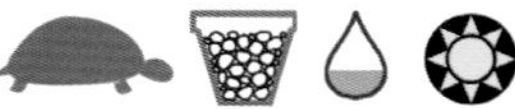

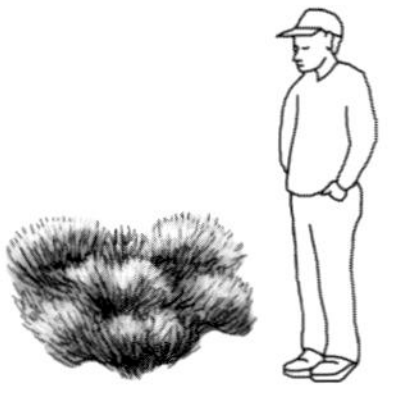

People often mistake green Mormon tea for Scotch broom, and it is broomlike in its upright habit and bright green, essentially leafless stems. But unlike Scotch broom, green Mormon tea is a well-behaved plant that never oversteps its bounds to become invasive. It is tolerant of a range of soil types, but it does appreciate good drainage. Green Mormon tea looks beautiful planted with yuccas, with soft green shrubs like big sagebrush and desert sage, and with Indian ricegrass, gooseberryleaf globemallow, and sundancer daisy. Once old enough to flower, an individual will produce either masses of bright yellow male flowers or tiny, dark brown cones, both of which are quite decorative.

*Special Features*: Green Mormon tea is a great asset in autumn and winter landscapes, as it keeps its structure and bright chartreuse green color year round.

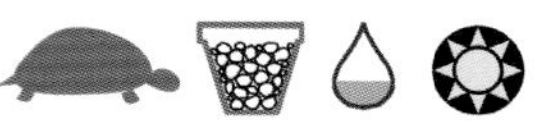

## Littleleaf Mountain Mahogany

***(Cercocarpus intricatus)***

This tough little shrub is usually found growing right out of the slickrock in the canyon country where it is most at home. Fortunately, it does not require such an extreme environment in order to thrive. It will be as happy in ordinary, well-drained garden soil as it is in a sandstone crack. It features shiny silver bark, almost needle-like, dark evergreen leaves, and a compact, intricately branched growth form. Its flowers are not very showy, but the feathery fruits that follow are quite attractive, especially when backlit. Littleleaf mahogany looks especially good planted with Fremont barberry and cliffrose. It can also be used effectively as a low hedge.

*Special Features*: Littleleaf mountain mahogany is quite tolerant of pruning and shaping. If you have deer, they may take care of this for you. Protecting new plantings is recommended.

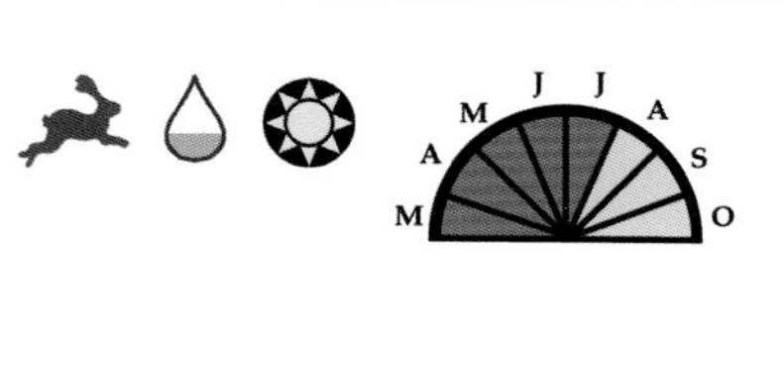

## Rubber Rabbitbrush

***(Ericameria nauseosa)***

A widely distributed and common roadside plant throughout our region, rubber rabbitbrush is often the first native shrub that newcomers learn to recognize and love. It features an absolutely glorious autumn display of rich golden flower masses, set off handsomely by the often nearly white, fine-textured foliage. Rubber rabbitbrush has a rubber-like scent that is pleasing to some and not so pleasing to others. It is broadly adapted and tough as nails, thriving in desert washes as well as in high mountain meadows. It will be happy as long as it gets plenty of sunshine. It looks stunning planted with bright green shrubs like fernbush and cliffrose, and it also combines well with larger perennials such as Palmer penstemon, prince's plume, and alkali sacaton grass.

*Special Features*: Once rubber rabbitbrush has finished flowering, it can be pruned back to the ground. This prevents massive volunteering from seed and keeps the plant compact and attractive. Pruning can also prolong its life. The plant will sprout back vigorously in the spring.

## Fernbush

***(Chamaebatiaria millefolium)***

Fernbush is named for its fine, fernlike foliage, which has a sweet, resinous scent. The plants are semi-evergreen, leafing out very early in the spring, and they have an open form that shows their intriguing leaf pattern to good advantage. Fernbush produces its showy spikes of cream-colored flowers in high summer, when most other ornamental trees and shrubs have long since finished flowering. The cinnamon-colored fruiting heads add interest into the fall. These should be clipped before the next flush of growth. Fernbush looks especially good planted with littleleaf mockorange, oakleaf sumac, and rubber rabbitbrush. It is not a fussy plant and will tolerate a broad range of conditions.

*Special Features*: Fernbush is dominant on the cinder fields at Idaho's Craters of the Moon, which shows you how tough this plant really is. But a little extra water will speed its growth.

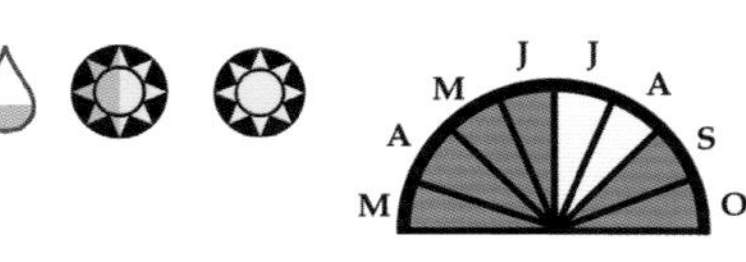

## Apache Plume

***(Fallugia paradoxa)***

Apache plume is primarily a plant of the washes of the desert Southwest. It enters our region via the Colorado Plateau and southern Great Basin. Do not be misled by these southern origins, however. Apache plume is a versatile and widely adapted plant, cold-hardy as far north as Saskatchewan. It has many attractive features, including an abundance of white flowers that resemble apple blossoms. These are followed by pink, feathery fruit clusters that last for weeks into the summer. The plant has a mounding growth form, satiny white bark, and small, deeply lobed green leaves. It combines well with other large, mounding shrubs such as oakleaf sumac and Fremont barberry in larger-scale landscapes, and it can also be used as a screen or specimen plant. It can be useful in minimal water landscapes if provided with harvested water.

*Special Features*: Apache plume is very tolerant of pruning, which may be necessary to scale back its tendency to spread by root sprouts. Oddly, it is rarely touched by deer.

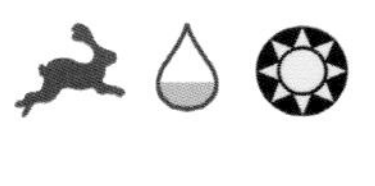

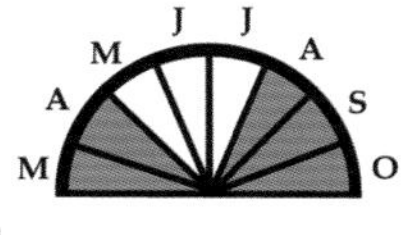

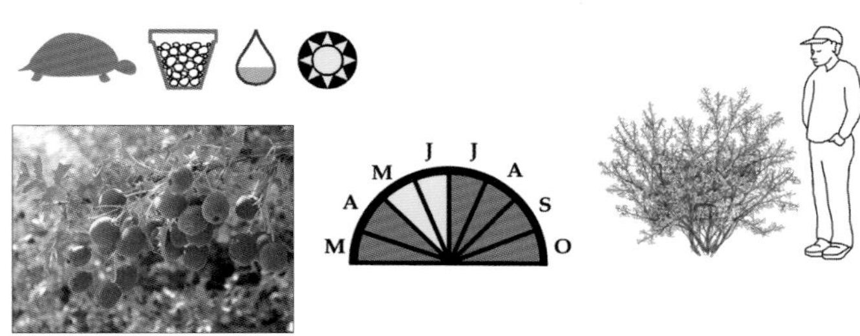


## Fremont Barberry

***(Mahonia fremontii)***

This handsome evergreen shrub is a characteristic species of the Colorado Plateau. Also called Utah holly, Fremont barberry has hard, holly-like leaves that are quite unfriendly to the touch. But they have a beautiful blue-green color, turning rose-purple in winter. Fremont barberry features masses of large, honey-scented golden blossoms in late spring. These are followed by fruits that are fleshy but hollow, like little balloons in party colors of purple and red. The fruits are edible and sweet, with a cluster of a few applelike seeds attached inside at the base. Fremont barberry is truly a plant for all seasons. It is a little slow to get started, but it is broadly adapted and not hard to grow, as long as the soil is well-drained and not kept too wet.

*Special Features*: With its sprawling growth form and deep blue-green foliage, Fremont barberry look great planted with shrubs that have contrasting forms and colors, for example, green Mormon tea, cliffrose, Apache plume, and rubber rabbitbrush.

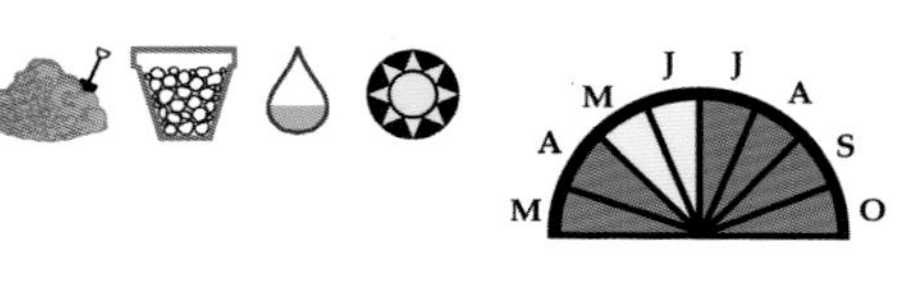

## Cliffrose

***(Purshia stansburiana)***

Cliffrose is among the most memorable shrubs of our region, and it has many excellent features. It is a tall, statuesque evergreen that has fine-textured, bright green foliage. It often takes on an interesting, rugged growth form as it matures. In early summer, cliffrose is graced with an abundance of pale yellow blossoms that have a spicy, clove-like fragrance. These are followed by feathery fruits that make the plant light up against a backdrop of late sun. Even the bark, which shreds off in tan and rosy strips, is beautiful. Cliffrose likes a coarse soil that is not very fertile, and it definitely does not like to have wet feet. It looks handsome planted with Fremont barberry and littleleaf mahogany. Deer love this plant, so give it protection when young.

*Special Features*: One of the most pleasing things about cliffrose is its scent. The foliage emits a lovely, resinous odor when warmed in the summer sun. Cliffrose wood also smells fragrant when it burns in a campfire.

## Creeping Oregon Grape

***(Mahonia repens)***

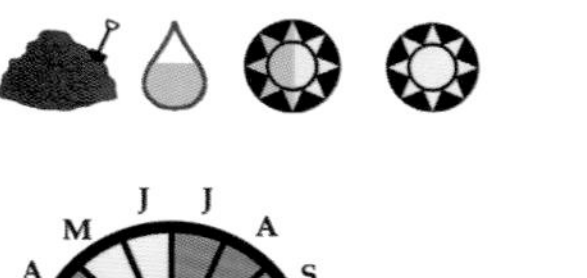

This pretty, evergreen, ground-cover plant can grow in sun or partial shade, but it tends to get taller in the sunshine, and it can winter-burn if the site is too bare and sunny. It has broad, leathery leaves with fine, spine-tipped teeth and fragrant sprays of golden flowers in spring. The clusters of grapelike berries are dark blue-violet with a waxy bloom. The berries are edible, and are attractive to birds. The leaves turn beautiful shades of purple and red in the winter. Best used as an understory plant beneath bigtooth maple, Gambel oak, or quaking aspen, it can also be used to stabilize a north- or east-facing slope. It combines well with mountain lover and common juniper.

*Special Features*: This plant is often confused with Oregon holly grape (*Mahonia aquifolium*), a larger plant of the Pacific Northwest, which is widely planted in our area and can be invasive. Our plant has dark green leaves with a "matte" finish, while the Northwestern species has bright green, lacquer-shiny leaves.

## Mountain Lover

***(Pachystima myrsinites)***

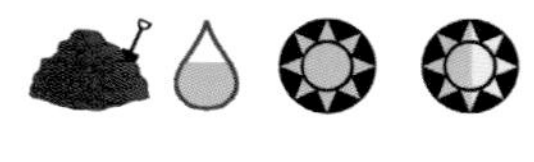

A graceful, fine-textured, evergreen ground cover plant, mountain lover is one of the few natives that prefers partial to full shade. It is well-named, as it rarely ventures out of the mountain forests that are its home. It likes rich soils, and, though it often grows under the relatively drought-tolerant little trees of the mountain brush community, it does not object to extra water. It can be quite successful as a landscape plant, as long as it has some protection from winter sun exposure, either under snow or in shade. Mountain lover has tiny flowers that are rarely noticed. It combines well with mountain perennials like columbines, as well as with other shade-tolerant ground cover plants.

*Special Features*: Unlike many native evergreens, mountain lover stays bright green all winter. It is startling to see its leafy branches springing up out of the snow on a warm January day.

# Woody Plants: Medium Water Use

**(Ranked Short to Tall)**

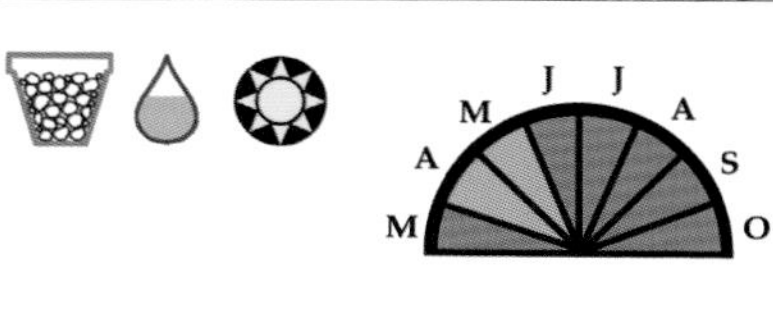

## Shrubby Penstemon

***(Penstemon fruticosus)***

Shrubby penstemon is widely distributed in the interior Northwest, and enters our region in eastern Washington and Oregon and central Idaho. It is one of the most reliable and beautiful native shrubs, featuring evergreen leaves that are a cheerful bright green and masses of large, spectacular, lavender flowers in mid-spring. Shrubby penstemon is a mountain plant, but it grows on steep, coarse scree slopes in full sun and is remarkably drought-hardy. Unlike many penstemons, it is a long-lived shrub that stays short and compact. A wonderful low hedge or bank planting, it can serve as a green backdrop for later-flowering species such as showy sandwort, sulfurflower buckwheat, and littlecup penstemon.

*Special Features*: Shrubby penstemon flowers are a magnet for hawk moths. These striped, furry insects are so large that they are easily mistaken for hummingbirds, especially in the dusk. The sight of these moths silently working the pale flowers in the twilight is truly magical.

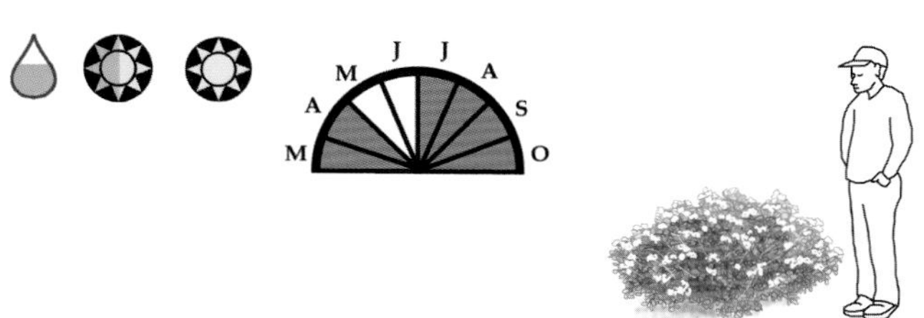


## Martin Mountain Lilac

***(Ceanothus martinii)***

One of many closely related species native to the West, Martin mountain lilac is a widely distributed but rarely encountered shrub of mountain brush communities in the southern Great Basin. It has a tidy, rounded growth form and coin-shaped green leaves that turn golden in the fall or sometimes persist all winter. But its main attraction is the veritable blizzard of fragrant white blossoms that cover the plant in spring. Each flower is tiny, but they are so densely packed on the branches that the leaves are often scarcely visible. Mountain lilac is an easygoing plant with no special needs. It combines well with mallowleaf ninebark and littleleaf mockorange.

*Special Features*: The seeds of mountain lilac are held in triplets in small, three-lobed capsules. When the capsule dries to a critical point, the seeds are catapulted away from the plant, making it hard to collect them.

## Woody Plants: Medium Water Use

**(Ranked Short to Tall)**

## Littleleaf Mockorange
***(Philadelphus microphyllus)***

This little-known shrub is quite similar to the mockorange species of wetter climes. It has the same pretty, four-petaled blossoms and the same memorable sweet scent. But its leaves are much finer, and the plant is much more drought-tolerant than cultivated mockorange species. It has a lovely, fountain-like form with slender, arching branches, and the satiny white bark peels in strips to reveal a cinnamon-colored underlayer. Its almond-shaped leaves turn a clear bright yellow in the fall. Littleleaf mockorange is not finicky and is easy to grow. It combines well with fernbush, green Mormon tea, and oakleaf sumac. Or try planting it along the edges of groves of bigtooth maple and Gambel oak.

*Special Features*: Plant littleleaf mockorange where you will be able to enjoy its spectacular floral display and beautiful fragrance on warm June nights.

## Oakleaf Sumac
***(Rhus trilobata)***

Oakleaf sumac is very widely distributed in western North America, and a similar species, fragrant sumac (*R. aromatica*), is found to the east. This plant has many fine features, including rapid growth, spectacular fall color in shades of red, orange, and yellow, bright red berries that attract birds, and the ability to tolerate a wide range of conditions. It does get quite large, so be sure to give it plenty of room. It can be used as a specimen plant, a screen or hedge, or to stabilize steep slopes. It is very tolerant of pruning, but trying to keep it small by pruning it back can be a losing battle. The common version has leaves with three leaflets. A somewhat smaller version with simple, scalloped leaves (*R. trilobata* var. *simplicifolia*) is found on drier sites in the southern half of its range, and this may be a better option for small landscapes.

*Special Features*: The red berries of oakleaf sumac are not edible, but they can be soaked in water to make a refreshing drink known locally as "boy scout lemonade."

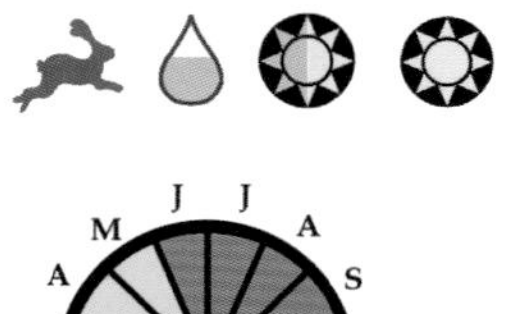

## Mallowleaf Ninebark

***(Physocarpus malvaceus)***

Mallowleaf ninebark is often a dominant plant in the mountain brush community at middle elevations. Its growth form varies by habitat—on steep, open slopes it stays short, while in more favorable bottoms and drainageways it grows much taller. Its size can thus be managed in the landscape with judicious watering. It has small, bright green, somewhat maplelike leaves, shiny red bark, and showy sprays of pink to white blossoms. It is an adaptable species, tolerant of a variety of conditions, but it does best in a rich soil with an organic mulch. It combines well with mountain lilac, littleleaf mockorange, and mountain snowberry.

*Special Features*: The leaves of mallowleaf ninebark turn an intense wine-red early in the fall, often coloring whole hillsides before the first hint of autumn color in any other species.

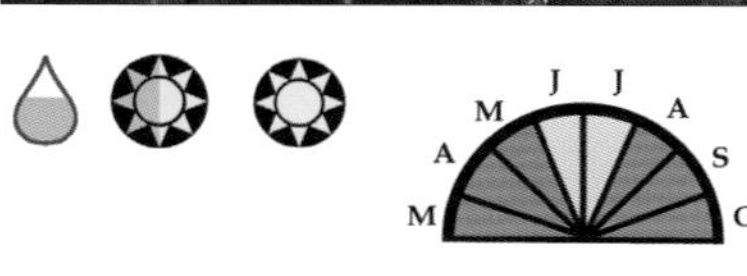


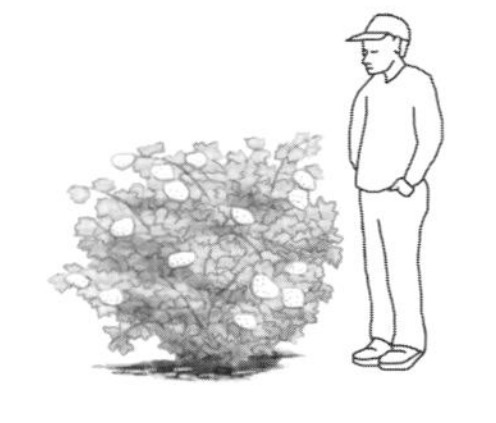

## Squaw Apple

***(Peraphyllum ramosissimum)***

Squaw apple combines narrow, bright green leaves with rather large, white to pink, delicately fragrant, apple-like blossoms. It is a relatively common member of mountain brush communities throughout the Intermountain region, but it is only occasionally seen in cultivation, perhaps because of its slow growth rate. It has an irregular form and smooth gray bark that becomes black and furrowed in older plants. Its pretty fruits resemble small, bright yellow and red apples. The leaves also turn bright yellow in the fall. Squaw apple is tolerant of a range of soil types and is not difficult to grow in the landscape. It is a favorite food of deer, however, so young plants need to be protected if deer browsing is an issue.

*Special Features*: The taste of squaw apple fruits is disappointing, to say the least, with a biting bitterness as the dominant element, though birds and squirrels take them readily, and they are known to be a favorite food of black bears.

## Alderleaf Mountain Mahogany

***(Cercocarpus montanus)***

A slender shrub that tends to form patches in the mountain brush zone, alderleaf mountain mahogany has smooth, pinkish gray bark and small, diamond-shaped leaves with deeply incised veins. In autumn the leaves turn a dark golden color, and the plant provides vertical structure in the winter garden. The feathery fruits add interest from middle to late summer, and are especially attractive when backlit. Alderleaf mountain mahogany is widely adapted and easy to grow. It makes a good addition to a mountain brush community planting, and can also be used effectively as a screen or specimen plant. It makes a good companion for fernbush, Utah serviceberry, and mountain lilac.

*Special Features*: During very dry summers in the wild, alderleaf mountain mahogany may lose its leaves. This somewhat alarming sight is an adaptation to avoid drought, and the plants will leaf out again normally the following spring.

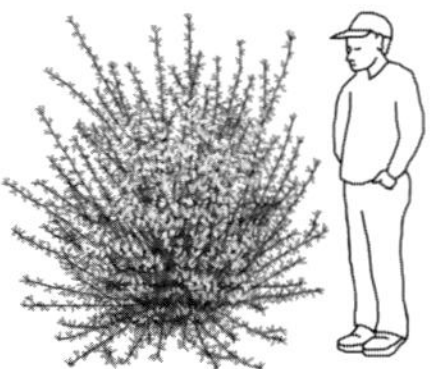

## Utah Serviceberry

***(Amelanchier utahensis)***

A graceful shrub or small tree with an open, vaselike form, Utah serviceberry is common and widely distributed throughout our region. It is found on drier sites in the mountain brush than its relative, Saskatoon serviceberry, which it closely resembles, and may occasionally be found in sagebrush steppe communities. Utah serviceberry features masses of white flowers in spring, followed by fruits that look like blueberries. The nearly round, soft, pale green leaves of serviceberry turn a beautiful dark gold in autumn, and its smooth gray bark and elegant form make a lovely sight in winter. Utah serviceberry is not picky about soils, though it likes an organic mulch. It looks good in an open planting with New Mexico privet and singleleaf ash. It is a popular deer food, so provide protection for young plants.

*Special Features*: The fruits of Utah serviceberry are dry, in contrast to the edible though rather flavorless fruits of Saskatoon serviceberry. Its chief advantages are greater drought hardiness and a somewhat showier flower display.

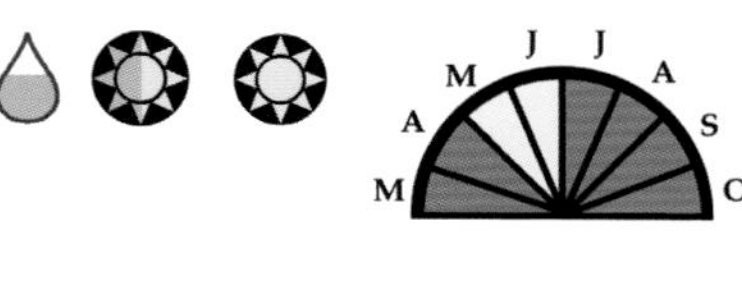

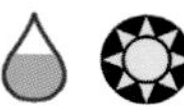

## New Mexico Privet

***(Forestiera pubescens)***

Also known as desert olive or stretchberry, New Mexico privet is a shrub or small tree of sandy floodplains on the Colorado Plateau. It is fully cold-hardy and has landscape potential far beyond its natural range of occurrence. It features smooth, cream to pale green bark on older wood, and striking black twigs. It has quite large, lime-green leaves, and clusters of showy purple fruits. Not every plant will produce fruits, as the sexes are on different individuals. New Mexico privet tends to be a shrub when grown in full sunlight, but in the gallery forests along rivers its form is more treelike. It can be limbed up to encourage this treelike form. This plant looks good with cliffrose, Utah serviceberry, and singleleaf ash, and can also be used in park-like settings with a native turf of blue grama.

*Special Features*: The olivelike fruits of New Mexico privet are popular with birds and also with coyotes in the wild. They can be a little messy in a landscape setting, though usually most are eaten before they have a chance to fall.

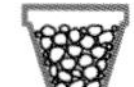

## Singleleaf Ash

***(Fraxinus anomala)***

This sturdy little tree is a common plant in the slickrock canyons of the Colorado Plateau. It features deep green, rounded, almost fleshy leaves, quite different from the slender compound leaves of other ash species. The lime-green flowers in spring are followed by clusters of dangling, canoe-paddle fruits that are quite attractive. Singleleaf ash also has striking black bark that is deeply furrowed on older branches. It often stays small and shrubby when growing out of rock, but in favorable microhabitats along washes it develops into a small tree. Singleleaf ash has been used very little in landscapes. In our experience, it is not difficult to grow, though it is slow to get started. It needs plenty of sunshine and space to prosper.

*Special Features*: The leaves of singleleaf ash turn a spectacular bright gold in autumn, and the black bark adds considerably to the dramatic effect.

## Curlleaf Mountain Mahogany

***(Cercocarpus ledifolius)***

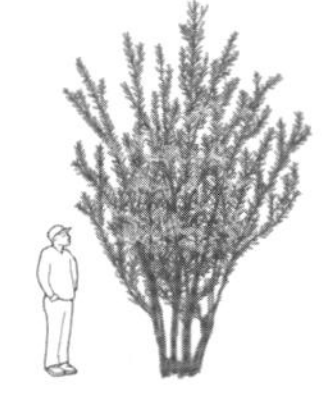

Curlleaf mountain mahogany is a tough little evergreen tree of windswept mountain ridges throughout our region. Its narrowly oval leaves are leathery and hard, with edges that are curled under, and they have a deep olive green color that contrasts nicely with the ridged gray bark. The trees are most conspicuous in fruit, when the feathery fruits can be so abundant as to nearly obscure the branches. These are especially noticeable when backlit. Curlleaf mountain mahogany is one of the best native trees for landscape use, because it stays small and has a pleasing, regular growth form. It makes a great specimen plant and is also very useful as a colonnade or a screen. This tree likes good drainage, but is otherwise not too particular about soil. It grows fairly quickly once it has had a year or two to get deeply rooted.

*Special Features*: Curlleaf mountain mahogany looks good in an open planting with other species from not-so-fertile environments, including datil yucca, Fremont barberry, cliffrose, and green Mormon tea.

## Western Virgin's Bower

***(Clematis ligusticifolia)***

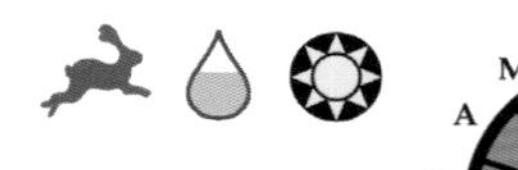

One of our few native woody vines, western virgin's bower makes an excellent choice to train over a pergola or arbor. It is a fast and vigorous grower and looks beautiful both when covered with clusters of cream-colored flowers and when these turn into feathery balls of seeds. It does not have tendrils or holdfasts, but instead climbs by twining its stems, as do domestic clematis species. It is densely leafy in summer, but these leaves are deciduous, so that much more light comes through in winter. The leaves themselves are bright lime green and have large, pointed leaflets that tend to be held vertically on the plant. They turn a clear yellow in autumn. Western virgin's bower is an easy plant to grow, but it requires pruning to direct its growth.

*Special Features*: Western virgin's bower looks quite a lot like oriental clematis. This introduced plant has become a nasty weed in many parts of our region. Oriental clematis has yellow, bell-shaped flowers that are solitary, not in clusters. Please do not plant it.

# Woody Plants: Medium Water Use
**(Ranked Short to Tall)**

## Rocky Mountain Juniper

***(Juniperus scopulorum)***

Rocky Mountain juniper is widely distributed and common throughout the West. Its slender, often weeping branchlets and small berries distinguish it from the native junipers of drier habitats. It commonly provides a strong evergreen element to mountain brush communities, and it is also found in aspen and ponderosa pine parkland at higher elevation, as well as in riparian areas of the shrub steppe. Rocky Mountain juniper has no special needs and is very easy to grow. This may be one reason why it has been so widely accepted into the landscape trade. Except possibly for blue spruce, it is planted more often than any other intermountain native. It combines well with Gambel oak, bigtooth maple, and oakleaf sumac in mountain brush plantings and also has many uses in more formal plantings.

*Special Features*: This plant is highly variable in form and color, and this has been the basis for the selection and release of numerous cultivars. We prefer species trees, but Rocky Mountain juniper cultivars certainly have their uses.

## Pinyon Pine

***(Pinus edulis)***

Pinyon pines are dominant species of foothill woodland communities throughout the southern half of our region, and they are largely responsible for its characteristic look and feel. The two-needle pinyon (*P. edulis*) is more common on the Colorado Plateau, while the single-needle pinyon (*P. monophylla*) is more common in the Great Basin. These familiar trees feature a twisted growth form, short, shiny green needles, and woody cones that contain edible pinenuts. Pinyon pines perform well in a landscape setting as long as they are not overwatered, though they can be slow-growing, especially the first few years. They can be used as specimen plants, in screen plantings, or as part of foothill woodland communities. They look great planted with green Mormon tea, datil and dwarf yucca, Fremont barberry, and cliffrose.

*Special Features*: The scent of pinyon pines in warm sun is one of the most familiar smells of the foothill woodlands.

## Gambel Oak
***(Quercus gambelii)***

Gambel oak is a common tree of mountain brush communities across the southern half of our region. It features furrowed, rough gray bark, a rugged, often arching growth form, and deeply lobed leaves with rounded lobes. These large, somewhat leathery leaves are among the last to fall in late autumn, after turning shades of deep gold, bronze, and red-purple. Its dramatic structure is accentuated by snow in the winter. Gambel oak is tolerant of a range of soil types, but prefers a rich but well-drained soil. It often grows with bigtooth maple and netleaf hackberry as its natural companions, along with understory species such as creeping Oregon grape, shining muttongrass, and rosy pussytoes.

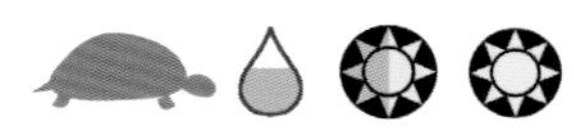

*Special Features*: Gambel oaks reproduce by suckering, often forming patches. In a landscape setting, the trees are usually so slow-growing that this suckering habit is not much of an issue. If you have Gambel oak on your building lot—build around it. Gambel oak does not transplant well, and the trees can take a long time to mature.

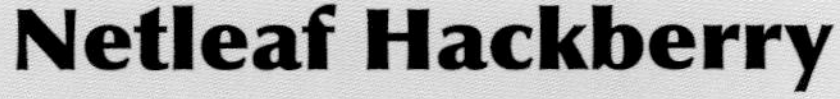

## Netleaf Hackberry
***(Celtis laevigata var. reticulata)***

This small tree is found throughout our region. It is important primarily in the mountain brush zone, but also grows in floodplain communities at lower elevations and occasionally as a minor component of the sagebrush steppe. It has a gnarled growth form, narrowly triangular, pointed leaves with deep veins and rough surfaces, and beautiful gray bark with corky ridges. Its small orange fruits have a thin but sweet pulp. They provide food for birds in winter. Netleaf hackberry is broadly adapted and occurs in a variety of soils. It is not difficult to grow. It can be used as a specimen, in screen or hedge plantings, or as a component of mountain brush community plantings. It will grow more quickly without an understory.

*Special Features*: One feature of this tree that you will love or hate is the tendency for the leaves to develop nipple galls, caused by a tiny, cicadalike insect. The silver-brown leaves with galls tend to cling to the tree through the winter, giving the effect of ghostly winter foliage.

## Bigtooth Maple

*(Acer grandidentatum)*

This elegant little tree is found almost exclusively in the Intermountain West. Its has a stately, fountainlike growth form when planted in the open under favorable conditions. It can be tall and aspenlike when growing in a forest setting, while on dry hillsides it forms shrubby thickets. It features smooth gray bark and classic Canada-flag maple leaves. Perhaps its most outstanding feature is its superb fall color, ranging from scarlet to deep crimson. Bigtooth maple is a common tree in the mountain brush zone and in lower-elevation streamside communities throughout the eastern half of our region. It is easy to grow, as long as it is grown as a species tree. Avoid bigtooth maple cultivars, as most are grown on sugar maple rootstocks—these rootstocks are not well adapted to western soils. Bigtooth maple can be used as a specimen plant, as a colonnade, or as part of a mountain brush community planting.

*Special Features*: Bigtooth maple is a close relative of sugar maple, and it can be tapped in the early spring for sap to make maple syrup.

## Common Juniper

***(Juniperus communis)***

This little evergreen occurs throughout the cooler regions of the Northern Hemisphere. It features a sprawling to prostrate growth form and spreading, shaggy branches cloaked with needlelike leaves that are dark or waxy blue green. It has reddish bark that peels off in shreds, and clusters of small, wrinkly, blue-purple berries. The plants root in at the branch nodes. Common juniper grows mostly on rocky soils in the mountains, but it is an adaptable plant that can grow in a range of soil types, as long as the soil is well-drained. This versatile plant looks good in formal foundation plantings, and the low forms are useful as a ground cover, especially for stabilizing steep slopes. Common juniper can also be used as an understory species for mountain forest plantings. It combines well with mountain lover and creeping Oregon grape.

*Special Features*: This plant has been in cultivation for a long time, and many cultivars are available. Most of these were developed from European collections.

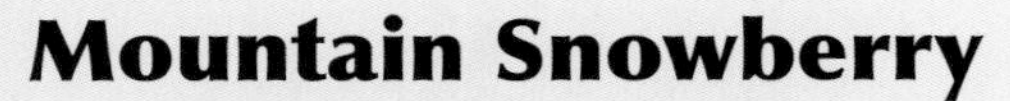

## Mountain Snowberry

***(Symphoricarpos oreophilus)***

Mountain snowberry is an attractive, fast-growing shrub that looks a lot like its relative common snowberry (*S. albus*), a plant that is much more common in cultivation, but mountain snowberry has longer, more tubular flowers. It is found throughout the West, and is often the dominant understory shrub in mountain brush, aspen parkland, and ponderosa pine communities. Its attractive features include a pretty, arching growth form, bright green leaves in pairs along the stems, pink, bell-shaped flowers, and soft, white berries. It can be used in formal settings, and also adds structure to mountain forest and meadow plantings. It prefers moist soils rich in organic matter. It combines well with squaw apple, golden currant, and mountain ninebark.

*Special Features*: The berries of mountain snowberry persist into the winter. It is a fine sight to see a flock of cedar waxwings descend onto snow-covered snowberry bushes and strip them of their berries in a matter of minutes.

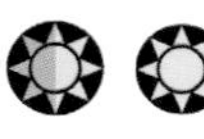

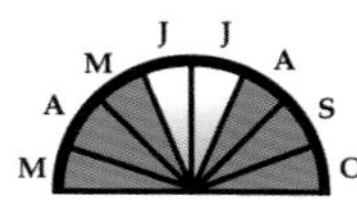

## Woody Plants: High Water Use

**(Ranked Short to Tall)**

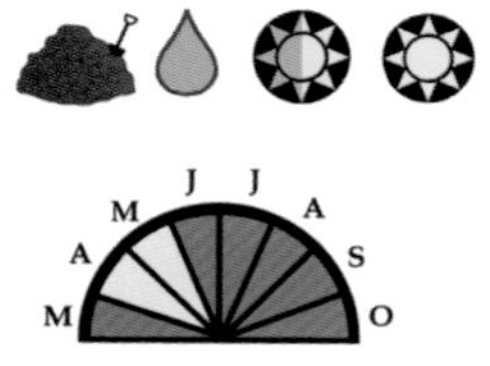

## Golden Currant

***(Ribes aureum)***

This handsome, tall shrub is one of the natives most familiar to the average gardener. It has many fine qualities, including beautiful, three-lobed leaves, fragrant yellow blossoms in late spring, fruits that are edible to both birds and humans, and outstanding red and orange fall color. The plants spread slowly by suckering and can be renewed by pruning out older stems. Golden currant prefers a rich, moist soil. It can be used as a specimen or screen, in foundation plantings, or in a mountain forest or meadow planting. It combines well with river birch, mountain snowberry, chokecherry, and western mountain ash. It is a great addition to a wildlife planting.

*Special Features*: In contrast with most intermountain natives, this plant is almost impossible to kill with too much water. This may explain its current popularity in home landscapes. It is no coincidence that virtually all native species currently used in intermountain home landscapes are plants of wet places.

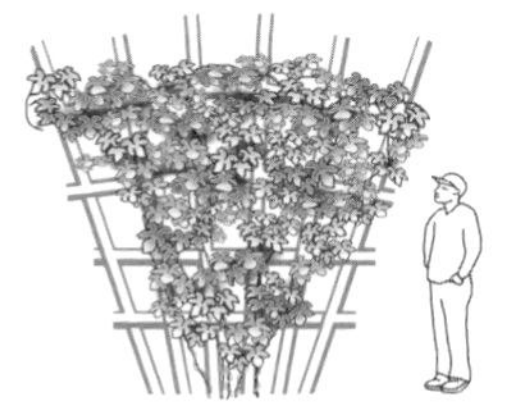

## American Hops

***(Humulus lupulus* var. *lupuloides)***

American hops is actually not a woody plant, because it dies back to the ground every winter and regenerates from the roots in the spring. We include it here because it can function in the landscape much as a woody vine does. It features rapid and vigorous growth each year, easily covering an arbor or pergola and creating an inviting, dappled shade effect. Its large, deeply lobed leaves have a tropical look, somewhat like mulberry leaves. They are rough to the touch and have deeply incised veins. The plant lacks tendrils or holdfasts for climbing, relying on its twining stems. The fruits look like little papery cones that hang down and are quite decorative. American hops is a streamside plant that requires rich soil and regular water in order to thrive. Its best use is to create shade over a ramada or arbor.

*Special Features*: American hops is a very close relative of the European hops used to flavor beer. The two species are now considered two varieties of a single species.

## Western Mountain Ash

***(Sorbus scopulina)***

Western mountain ash is related to the European rowan (*S. aucuparia*). It features smooth, almost shiny, red-brown bark, large, compound leaves, sprays of white flowers in late spring, and red-orange berries that are very ornamental. Western mountain ash is widely distributed in our region, usually as an understory tree in aspen and white fir communities or at streamside. It is sometimes found in mountain brush communities on moist northern slopes. Western mountain ash prefers rich soil and regular water, making it readily adaptable to traditional gardens. It can be used as a specimen plant or screen, or planted with taller trees for a charming multi-story effect. It combines well with quaking aspen, chokecherry, golden currant, and river hawthorn.

*Special Features*: This little tree really comes into its own in autumn. The combination of beautifully displayed leaves that turn magnificent shades of gold, orange, and red with sprays of crayon-orange berries is unforgettable.

## River Hawthorn

***(Crataegus rivularis)***

This handsome little tree is usually found on floodplains and at streamside. It features an upright growth form, usually with a single main trunk, and with branches that are at right angles to the trunk. It has shiny toothed leaves that turn yellow in fall, sprays of cherrylike blossoms, and deep purple fruits that resemble small apples. The fruits are tasty though rather seedy. They are readily taken by birds. This plant prefers rich soil and regular water. River hawthorn has a tendency to spread by rootstocks, but this tendency can be controlled by targeted watering. This plant can be used as a specimen or screen, or as part of a multistoried mountain forest planting. It combines well with golden currant, western mountain ash, chokecherry, and river birch.

*Special Features*: Like most hawthorns, river hawthorn is heavily armed with stout thorns, making it a great choice as a bird haven where there are cats or other climbing predators. It combines nesting sites with a built-in food supply.

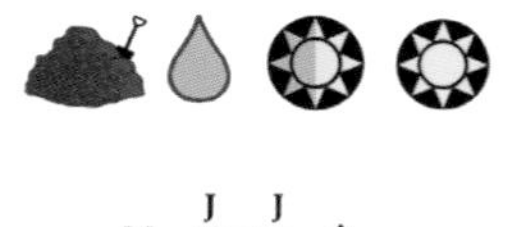

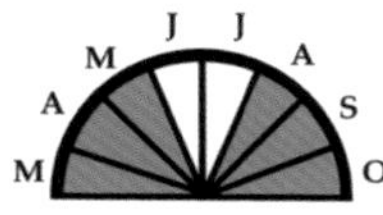

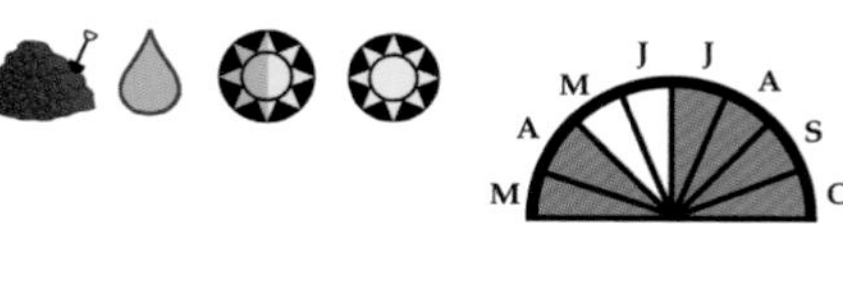


## Chokecherry
***(Prunus virginiana)***

Chokecherry is familiar as the source of fruits for delicious syrups and jellies. It is common in streamside and aspen communities throughout North America, and is also found in mountain brush communities on cooler north slopes. Chokecherry is a small, multitrunked tree with finely toothed cherry leaves and bright red twigs. The older bark is ashy gray and deeply furrowed. Its cream-colored flowers are borne in dense spikes that hang downward and have a pleasant, almondlike fragrance. They are followed by loose clusters of small, dark red cherries that have a sweet if astringent flavor. Chokecherry prefers rich soils and regular watering. Best used as a specimen plant or as part of a multi-storied mountain planting, it combines well with river hawthorn, Rocky Mountain maple, western mountain ash, and quaking aspen.

*Special Features*: Chokecherry is a good plant to include in a wildlife planting, as its flowers are visited by many interesting pollinators and its juicy fruits are quite attractive to birds.

## Great Basin Bristlecone Pine
***(Pinus longaeva)***

This slow-growing tree of the high mountains makes a surprisingly good landscape tree, especially where there is a need for a pine that will not outgrow its space. It is full of character even as a young tree, with an irregular crown form and short, stiff needles that cloak the branches, giving a bottlebrush effect. This character only increases as the tree grows older. Bristlecone pine is best used as a specimen plant, so that it will have room to develop its interesting shape. It needs a well-drained, preferably rocky soil to prosper. It could be used as the centerpiece of a rock garden or in combination with other conifers. It also looks good planted with flowering shrubs such as squaw apple or littleleaf mockorange.

*Special Features*: Plants of this species growing at timberline in mountains of the Great Basin are the oldest known living trees, some as old as 4,700 years. It will grow more quickly in a garden setting than on the ridgetops where these ancients are found.

## Rocky Mountain Maple

***(Acer glabrum)***

Rocky Mountain maple is a slender, elegant, often multitrunked tree that is widely distributed in our region. It is usually found along streams in the foothills, but is associated with ponderosa pine, aspen, and white fir at higher elevations. The leaves of Rocky Mountain maple have three pointed lobes and coarse to fine teeth along the edges. They usually turn a clear yellow in autumn. This plant likes rich soil and moist conditions. It works well as a specimen or screen, or as part of a multistoried mountain forest planting. It is quite shade tolerant and can coexist with evergreens like white fir, which set it off to great advantage, especially in fall.

*Special Features*: Rocky Mountain maple features bright red bark on new growth, somewhat like the bark of red osier dogwood. The bark on older wood is gray and slightly roughened.

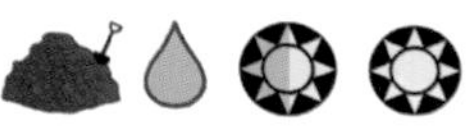

## Western Water Birch

***(Betula occidentalis)***

This fast-growing, multitrunked tree is usually slender and graceful, but it can get quite massive with age. It is found throughout the West, usually along streams and floodplains, but sometimes on north-facing mountain slopes. River birch features a fountain-like growth form and bright green, rather leathery, finely toothed leaves. These put on a fine display of gold to bronze fall color. River birch is more drought hardy than its name implies, though it prefers moist, rich soil. It is tolerant of a range of soil types, including heavy clay soils. It makes an excellent specimen plant and can also be used as part of a mountain forest community. It combines well with golden currant, river hawthorn, and Rocky Mountain maple.

*Special Features*: The most striking feature of river birch is its smooth, shiny red bark accented with horizontal white stripes (lenticels). This bark is reminiscent of cherry bark.

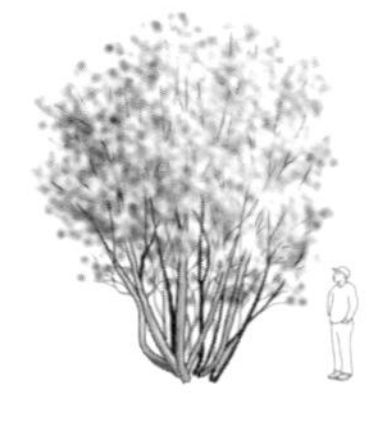

## Quaking Aspen

***(Populus tremuloides)***

Probably the best-loved native tree of the region, quaking aspen is also one of the most problematic in cultivation. We do not recommend quaking aspen for desert or semi-desert sites. It sprouts vigorously, especially when one of many insect and disease ailments attacks the parent tree. Keeping trees healthy is the best defense, but this is much easier in a mountain environment. Planting on a north-facing exposure, using a soil amended with plenty of organic matter, and using a thick organic mulch can improve success in the foothills. Targeting the application of water also helps control sprouting--the over-watered lawn is usually the first area to turn into a sprout forest. Appropriately placed, aspen is a beautiful tree that looks great planted with white fir and Rocky Mountain juniper.

*Special Features*: This is the tree *par excellence* for golden fall color; the contrast of shimmering golden leaves with creamy white bark and black branches is truly stunning.

## White Fir

***(Abies concolor)***

White fir is a beautiful evergreen tree of the intermountain region, and it deserves much wider use. It has a regular, conical shape when young but develops a more rounded crown with age. The blue-green, flattened needles are curved upward from the twigs, giving the foliage a soft look and feel. Metallic red male cones and large, cylindrical, pale green to purplish female cones add interest on older trees. White fir grows well in a range of soil types, but will grow more quickly in a silt or loam soil. It looks great planted with Rocky Mountain and bigtooth maples, river birch, and aspen, and also makes a pretty backdrop for flowering shrubs such as mallowleaf ninebark and mountain lilac.

*Special Features*: White fir has a root system that is both deep and spreading, making it more drought hardy than many more widely used conifers and also less subject to wind throw in the sometimes violent summer storms that characterize our region.

## Ponderosa Pine

***(Pinus ponderosa)***

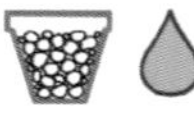

This majestic tree is common throughout the West, but it is often overlooked for landscape use. It features deeply plated red and yellow bark, long, shining needles, and a predictable, upright growth form. It is a large tree, but correctly placed, it can be a great shade tree, much better than most conifers. Because it has a natural tendency to self-prune as it gets taller, it looks quite natural when limbed up. This makes it easy to create a shady living space beneath its canopy, with soft, springy pinestraw for a floor. Ponderosa pine is broadly adapted and not difficult to grow. It is a natural companion for understory species like Indian ricegrass, little bluestem, and sulfurflower buckwheat. It also combines well with fernbush, mountain lilac, and alderleaf mountain mahogany.

*Special Features*: One of the most attractive things about ponderosa pine is its smell. It has a fresh, piney scent even from a distance. But if you get close enough to stick your nose in the bark fissures, you will be rewarded with a spicy vanilla fragrance.

## Table of Perennial Plants
### Ranked by Water Zone and Height

| Page | Scientific Name | Common Name | Family | Ht. | Crn. | Water | Light | Drain | OM |
|---|---|---|---|---|---|---|---|---|---|
| 166 | *Sphaeralcea caespitosa* | Cushion Globemallow | Malvaceae | 0.25 | 1 | min | sun | yes | |
| 166 | *Oenothera caespitosa* | Fragrant Evening Primrose | Onagraceae | 0.5 | 1 | min | sun | yes | |
| 167 | *Castilleja angustifolia dubia* | Indian Paintbrush | Scrophulariaceae | 1 | 1 | min | sun | yes | |
| 167 | *Eriogonum ovalifolium* | Silver Buckwheat | Polygonaceae | 1 | 0.5 | min | sun | yes | |
| 168 | *Tetraneuris acaulis* | Sundancer Daisy | Asteraceae | 1 | 0.5 | min | sun | yes | |
| 168 | *Oxytropis lambertii* | Purple Crazypea | Fabaceae | 1 | 0.5 | min | sun | yes | |
| 169 | *Penstemon utahensis* | Utah Penstemon | Scrophulariaceae | 1.5 | 1 | min | sun | yes | low |
| 169 | *Arenaria macradenia* | Showy Sandwort | Caryophyllaceae | 1.5 | 2 | min | sun | yes | |
| 170 | *Achnatherum hymenoides* | Indian Ricegrass | Poaceae | 1.5 | 1.5 | min | sun | yes | |
| 170 | *Sphaeralcea grossulariifolia* | Gooseberryleaf Globemallow | Malvaceae | 2 | 2 | min | sun | | |
| 171 | *Achnatherum speciosum* | Desert Needlegrass | Poaceae | 2 | 2 | min | sun | | |
| 171 | *Stanleya pinnata* | Prince's Plume | Brassicaceae | 3 | 3 | min | sun | yes | low |
| 172 | *Penstemon palmeri* | Palmer Penstemon | Scrophulariaceae | 5 | 3 | min | sun | yes | low |
| 173 | *Astragalus utahensis* | Utah Ladyfinger Milkvetch | Fabaceae | 0.5 | 1 | low | sun | yes | low |
| 173 | *Stenotus acaulis* | Dwarf Goldenbush | Asteracerae | 0.5 | 1 | low | sun | yes | low |
| 174 | *Eriogonum brevicaule* | Shortstem Buckwheat | Polygonaceae | 0.75 | 1 | low | sun | | |
| 174 | *Eriogonum umbellatum* | Sulfurflower Buckwheat | Polygonaceae | 1 | 1 | low | sun | | |
| 175 | *Erigeron argentatus* | Silver Daisy | Asteraceae | 1 | 1 | low | sun | yes | |
| 175 | *Bouteloua gracilis* | Blue Grama | Poaceae | 1.25 | 1 | low | sun | | |
| 176 | *Poa fendleriana* | Shining Muttongrass | Poaceae | 1.25 | 1 | low | sun/part | | |
| 176 | *Hedysarum boreale* | Utah Sweetvetch | Fabaceae | 1.5 | 1 | low | sun/part | | |
| 177 | *Mirabilis multiflora* | Desert Four O'Clock | Nyctaginaceae | 1.5 | 5 | low | sun | yes | |
| 177 | *Gaillardia pinnatifida* | Hopi Blanketflower | Asteraceae | 2 | 1 | low | sun | | |
| 178 | *Linum lewisii* | Lewis Flax | Linaceae | 2 | 2 | low | sun | | |
| 178 | *Penstemon cyananthus* | Wasatch Penstemon | Scrophulariaceae | 2 | 1 | low | sun/part | yes | |
| 179 | *Pseudoroegneria spicata* | Bluebunch Wheatgrass | Poaceae | 2 | 1.5 | low | sun | | |
| 179 | *Penstemon rostriflorus* | Bridges Penstemon | Scrophulariaceae | 2 | 2.5 | low | sun | yes | |
| 180 | *Penstemon sepalulus* | Littlecup Penstemon | Scrophulariaceae | 3 | 2 | low | sun | yes | |
| 180 | *Penstemon eatonii* | Firecracker Penstemon | Scrophulariaceae | 3 | 2 | low | sun | yes | low |
| 181 | *Sporobolus airoides* | Alkali Sacaton Grass | Poaceae | 3 | 3 | low | sun | | |
| 182 | *Penstemon caespitosus* | Mat Penstemon | Scrophulariaceae | 0.25 | 1* | med | sun | | |
| 182 | *Antennaria rosea* | Rosy Pussytoes | Asteraceae | 0.5 | 1* | med | sun/part | | |
| 183 | *Erigeron flagellaris* | Trailing Daisy | Asteraceae | 0.5 | 1* | med | sun/part | | high |
| 183 | *Penstemon leonardii* | Leo Penstemon | Scrophulariaceae | 0.5 | 1 | med | sun | yes | |
| 184 | *Calylophus lavandulifolius* | Lavenderleaf Sundrops | Onagraceae | 0.5 | 1 | med | sun | yes | |
| 184 | *Penstemon linarioides* | Flaxleaf Penstemon | Scrophulariaceae | 1 | 1 | med | sun | yes | |
| 185 | *Asclepias tuberosa* | Butterfly Milkweed | Asclepiadaceae | 1.5 | 2 | med | sun | | high |
| 185 | *Penstemon whippleanus* | Whipple Penstemon | Scrophulariaceae | 2 | 1 | med | sun/part | yes | high |
| 186 | *Schizachyrium scoparium* | Little Bluestem | Poaceae | 2 | 1.5 | med | sun | | |
| 186 | *Penstemon comarrhenus* | Dusty Penstemon | Scrophulariaceae | 3 | 1 | med | sun/part | yes | |
| 187 | *Iliamna rivularis* | Maple Mallow | Malvaceae | 5 | 3 | med | sun/part | | high |
| 187 | *Leymus cinereus* | Basin Wildrye | Poaceae | 6 | 3 | med | sun | | |
| 188 | *Geum triflorum* | Prairie Smoke | Rosaceae | 0.75 | 1.5 | high | sun | | |
| 188 | *Monardella odoratissima* | Little Beebalm | Lamiaceae | 1 | 1.5 | high | sun | yes | |
| 189 | *Lithospermum multiflorum* | Mountain Puccoon | Boraginaceae | 1.5 | 1.5 | high | sun/part | | |
| 189 | *Epilobium canum* | Firechalice | Onagraceae | 1.5 | 1.5 | high | sun/part | | |
| 190 | *Aquilegia coerulea* | Rocky Mountain Columbine | Ranunculaceae | 2 | 1.5 | high | sun/part | | high |
| 190 | *Aquilegia formosa* | Western Columbine | Ranunculaceae | 2 | 1.5 | high | sun/part | | high |
| 191 | *Erigeron speciosus* | Showy Daisy | Asteraceae | 2 | 2 | high | sun/part | | |
| 191 | *Hymenoxys hoopesii* | Meadow Fire | Asteraceae | 2 | 2 | high | sun/part | | |
| 192 | *Geranium viscosissimum* | Sticky Geranium | Geraniaceae | 3 | 3 | high | sun/part | | |
| 192 | *Chamerion angustifolium* | Blooming Sally | Onagraceae | 3 | 1.5* | high | sun/part | | |
| 193 | *Delphinium barbeyi* | Tall Larkspur | Ranunculaceae | 3 | 3 | high | sun/part | | high |

* Plants spread and form bigger patches.
Ht. = height in feet
Crn = crown diameter in feet
Water = water zone: min (minimal), low, med (medium), high

Light = light requirement: sun, part (partial shade), shd (full shade)
Drain = drainage: yes (exceptionally good drainage needed)
OM=organic matter: low (shuns organic matter), high (prefers organic matter)

# Table of Woody Plants (Trees, Shrubs, and Subshrubs)

**Ranked by Water Zone and Height.**

| Page | Scientific Name | Common Name | Family | Ht. | Crn. | Water | Light | Drain | OM |
|---|---|---|---|---|---|---|---|---|---|
| 194 | *Atriplex confertifolia* | Shadscale | Chenopodiaceae | 1.5 | 2 | min | sun | yes | low |
| 194 | *Krascheninnikovia lanata* | Winterfat | Chenopodiaceae | 2.5 | 2 | min | sun | yes | low |
| 195 | *Eriogonum corymbosum* | Lacy Buckwheatbrush | Polygonaceae | 2.5 | 3 | min | sun | | |
| 195 | *Salvia dorrii* | Desert Sage | Lamiaceae | 2.5 | 3 | min | sun | yes | low |
| 196 | *Yucca baccata* | Datil Yucca | Agavaceae | 3 | 3 | min | sun | yes | low |
| 196 | *Yucca harrimaniae* | Dwarf Yucca | Agavaceae | 3 | 3 | min | sun | yes | low |
| 197 | *Artemisia filifolia* | Sand Sagebrush | Asteraceae | 4 | 4 | min | sun | yes | low |
| 198 | *Artemisia tridentata* | Big Sagebrush | Asteraceae | 3 | 3 | low | sun | | |
| 198 | *Ephedra viridis* | Green Mormon Tea | Ephedraceae | 3 | 4 | low | sun | yes | |
| 199 | *Cercocarpus intricatus* | Littleleaf Mountain Mahogany | Rosaceae | 3 | 4 | low | sun | yes | |
| 199 | *Ericameria nauseosa* | Rubber Rabbitbrush | Asteraceae | 4 | 4 | low | sun | | |
| 200 | *Chamaebatiaria millefolium* | Fernbush | Rosaceae | 5 | 5 | low | sun/part | | |
| 200 | *Fallugia paradoxa* | Apache Plume | Rosaceae | 5 | 5 | low | sun | | |
| 201 | *Mahonia fremontii* | Fremont Barberry | Berberidaceae | 6 | 6 | low | sun | yes | |
| 201 | *Purshia stansburiana* | Cliffrose | Rosaceae | 10 | 4 | low | sun | yes | low |
| 202 | *Mahonia repens* | Creeping Oregon Grape | Berberidaceae | 1 | 3* | med | sun/part | | high |
| 202 | *Pachystima myrsinites* | Mountain Lover | Buxaceae | 1 | 4 | med | part/shd | | high |
| 203 | *Penstemon fruticosus* | Shrubby Penstemon | Scrophulariaceae | 2 | 2 | med | sun | yes | |
| 203 | *Ceanothus martinii* | Martin Mountain Lilac | Rhamnaceae | 3 | 5 | med | sun/part | | |
| 204 | *Philadelphus microphyllus* | Littleleaf Mockorange | Saxifragaceae | 4 | 5 | med | sun/part | yes | |
| 204 | *Rhus trilobata* | Oakleaf Sumac | Anacardiaceae | 4 | 6 | med | sun/part | | |
| 205 | *Physocarpus malvaceus* | Mallowleaf Ninebark | Rosaceae | 4 | 4 | med | sun/part | | |
| 205 | *Peraphyllum ramosissimum* | Squaw Apple | Rosaceae | 5 | 5 | med | sun | | |
| 206 | *Cercocarpus montanus* | Alderleaf Mountain Mahogany | Rosaceae | 6 | 5 | med | sun/part | | |
| 206 | *Amelanchier utahensis* | Utah Serviceberry | Rosaceae | 10 | 6 | med | sun/part | | |
| 207 | *Forestiera pubescens* | New Mexico Privet | Oleaceae | 10 | 8 | med | sun | | |
| 207 | *Fraxinus anomala* | Singleleaf Ash | Oleaceae | 12 | 12 | med | sun | yes | |
| 208 | *Cercocarpus ledifolius* | Curlleaf Mountain Mahogany | Rosaceae | 15 | 15 | med | sun | yes | |
| 208 | *Clematis ligusticifolia* | Western Virgin's Bower | Ranunculaceae | 15 | 15 | med | sun | | |
| 209 | *Juniperus scopulorum* | Rocky Mountain Juniper | Cupressaceae | 20 | 12 | med | sun/part | | |
| 209 | *Pinus edulis* | Pinyon Pine | Pinaceae | 20 | 15 | med | sun | yes | |
| 210 | *Quercus gambelii* | Gambel Oak | Fagaceae | 20 | 15 | med | sun/part | | |
| 210 | *Celtis laevigata reticulata* | Netleaf Hackberry | Ulmaceae | 20 | 20 | med | sun/part | | |
| 211 | *Acer grandidentatum* | Bigtooth Maple | Aceraceae | 30 | 25 | med | sun/part | | |
| 212 | *Juniperus communis* | Common Juniper | Cupressaceae | 3 | 6 | high | part/shd | | |
| 212 | *Symphoricarpos oreophilus* | Mountain Snowberry | Caprifoliaceae | 3 | 4 | high | sun/part | | high |
| 213 | *Ribes aureum* | Golden Currant | Saxifragaceae | 6 | 6 | high | sun/part | | high |
| 213 | *Humulus lupulus lupuloides* | American Hops | Moraceae | 10 | 5 | high | sun/part | | high |
| 214 | *Sorbus scopulina* | Western Mountain Ash | Rosaceae | 10 | 8 | high | sun/part | | high |
| 214 | *Crataegus rivularis* | River Hawthorn | Rosaceae | 12 | 10 | high | sun/part | | high |
| 215 | *Prunus virginiana* | Chokecherry | Rosaceae | 15 | 10 | high | sun/part | | high |
| 215 | *Pinus longaeva* | Great Basin Bristlecone Pine | Pinaceae | 20 | 15 | high | sun | yes | |
| 216 | *Acer glabrum* | Rocky Mountain Maple | Aceraceae | 20 | 10 | high | sun/part | | high |
| 216 | *Betula occidentalis* | Western Water Birch | Betulaceae | 20 | 15 | high | sun/part | | high |
| 217 | *Populus tremuloides* | Quaking Aspen | Salicaceae | 30 | 20 | high | sun | | high |
| 217 | *Abies concolor* | White Fir | Pinaceae | 70 | 20 | high | sun/part | yes | high |
| 218 | *Pinus ponderosa* | Ponderosa Pine | Pinaceae | 70 | 20 | high | sun | yes | |

* Plants spread and form bigger patches.
Ht. = height in feet
Crn = crown diameter in feet
Water = water zone: min (minimal), low, med (medium), high

Light = light requirement: sun, part (partial shade), shd (full shade)
Drain = drainage: yes (exceptionally good drainage needed)
OM=organic matter: low (shuns organic matter), high (prefers organic matter)

# Plant Names Index

Scientific and common names we use in the Plant Palette are shown in this index in **bold**. Scientific names are shown in *italics*. We refer to the USDA Plants Database (plants.usda.gov) as our authority for current scientific names. Because of recent advances in plant taxonomy, quite a few names have changed, and the new names may be unfamiliar to many people. We have also included the old, familiar scientific names for several of these plants, along with their modern versions. Not all of the old names are simple synonyms for the new names—sometimes the story is more complex. Similarly, many plants have multiple common names—the name we prefer to use is listed in bold, but other names are listed as well, along with the equivalent name used here.

*Cushion globemallow*

# Resources for Further Information

Appendix

## Native Landscapes of the Intermountain West

### State Native Plant Societies

A good way to learn more about the native plants and plant communities in your area is to contact the local native plant society. These folks sponsor interesting and informative field trips to wild places, plant identification workshops, and other great activities. Often just a trip to a native plant society web site can be very educational, and a chance to go plant watching with someone who already knows the plants can be invaluable.

Arizona Native Plant Society (5 chapters)
520–882–7663; anps@aznps.org
www.aznps.org/

California Native Plant Society
(34 chapters)
916–447–2677; cnps@cnps.org
www.cnps.org/

Southern California Botanists
714–448–7034;
aromspert@fullerton.edu
www.socalbot.org/

Colorado Native Plant Society
(6 chapters)
303–556–8309;
kimberly.regier@cudenver.edu
www.conps.org/

Idaho Native Plant Society (7 chapters)
208–342–2631; idahoplants@arczip.com
www.IdahoNativePlants.org/

Montana Native Plant Society (7 chapters)
406–542–0202; susan213@msn.com
www.umt.edu/mnps/

Nevada Native Plant Society
702–329–1645; atiehm@unr.edu
heritage.nv.gov/nnps.htm/

Native Plant Society of New Mexico (8 chapters)
505–523–8413; jfreyerm@lib.nmsu.edu
npsnm.unm.edu/

Native Plant Society of Oregon (13 chapters)
503–236–8787; president@NPSOregon.org
www.npsoregon.org/

Utah Native Plant Society (8 chapters)
801–377–5918; unps@unps.org
www.unps.org/index.html

Washington Native Plant Society (11 chapters)
206–527–3210; wnps@wnps.org
www.wnps.org/

Wyoming Native Plant Society (2 chapters)
605–673–3159
uwadmnweb.uwyo.edu/WYNDD/wnps/wnps_home.htm

North American Native Plant Society
www.nanps.org/

## Botanical Gardens and Arboreta

Several botanical gardens and arboreta in our region feature sections, displays, and information on native plant horticulture.

ARIZONA
The Arboretum at Flagstaff
4001 Woody Mountain Road
Flagstaff, AZ 86001–8775
928–774–1442; www.thearb.org/

Arizona–Sonora Desert Museum
2021 N. Kinney Road
Tucson, AZ 85743–8918
520–883–1380; www.desertmuseum.org/

Desert Botanical Garden
1201 North Galvin Pkwy.
Phoenix, AZ 85008
480–941–1225; www.dbg.org/

CALIFORNIA
Davis Arboretum
University of California
One Shields Avenue
Davis, CA 95616–8526
530–752–4880; arboretum.ucdavis.edu/

Quail Botanical Gardens
230 Quail Gardens Drive
Encinitas CA 92024
760–436–3036; www.qbgardens.com/

Rancho Santa Ana Botanic Garden
1500 North College Ave.
Claremont, CA 91711–3157
909–625–8767; www.rsabg.org/

Santa Barbara Botanic Garden
1212 Mission Canyon Road
Santa Barbara, CA 93105
805–682–4726; www.sbbg.org/

Strybing Arboretum and Botanical Gardens
9th Avenue & Lincoln Way
San Francisco, CA 94122
415–661–1316; www.strybing.org/

University of California Botanical Garden
200 Centennial Dr. #5045
Berkeley, CA 94720–5045
510–642–0849;
botanicalgarden.berkeley.edu/

COLORADO
Denver Botanic Gardens
909 York St.
Denver, CO 80206
720–865–3500; www.botanicgardens.org/

Western Colorado Botanical Gardens and Butterfly House
641 Struthers Avenue
Grand Junction, CO 81501
970–245–3288; www.wcbotanic.org/

IDAHO
Idaho Botanical Garden
2355 N. Penitentiary Road
Boise, ID 84712
208–343–8649; 877–527–8233;
www.idahobotanicalgarden.org

Sawtooth Botanical Garden
P.O. Box 928, Sun Valley, ID, 83353
208–726–9358; www.sbgarden.org/

NEVADA
Wilbur D. May Arboretum and Botanical Garden
1595 N. Sierra St.
Reno, NV 89503
775–785–4153; maycenter.com/

NEW MEXICO
Rio Grand Botanic Garden
903 10th St. SW
Albuquerque, NM 87102
505–764–6200;
www.cabq.gov/biopark/garden/index.html

OREGON
Berry Botanic Garden
11505 SW Summerville Ave.
Portland, OR 97219–8309
503–636–4112; www.berrybot.org/

UTAH
Red Butte Garden and Arboretum
300 Wakara Way
Salt Lake City, UT 84108
801–581–4747; www.redbuttegarden.org/

Utah State University
Utah Botanical Center Home
725 Sego Lily Drive
Kaysville, UT 84037
801–593–8969
http://utahbotanicalcenter.org/

WASHINGTON
Bellevue Botanical Garden
12001 Main Street
Bellevue, WA 98005
425–452–2750;
www.bellevuebotanical.org/

## Plant Ecology and Identification Books

There are many good popular field guides to the plants of different parts of our region. Some of these books are recent, some old, but even the older ones are often available from online booksellers. All those listed feature plant pictures for help with identification. We also list a few books that discuss the plants in the larger context of the natural history of the region.

### *Great Basin*

Blackwell, Laird R. *Great Basin Wildflowers: A Guide to Common Wildflowers of the High Deserts of Nevada, Utah, and Oregon*. Guilford, CT: Falcon Press, 2006.

Blackwell, Laird R. *Wildflowers of the Eastern Sierra and Adjoining Mojave Desert and Great Basin*. Edmonton, AB: Lone Pine Publishing, 2002.

Lanner, Ronald M. *Trees of the Great Basin*. Reno, NV: University of Nevada Press, 1984.

Mozingo, Hugh N. *Shrubs of the Great Basin*. Reno, NV: University of Nevada Press, 1987.

Taylor, Ronald, J. *Sagebrush Country: A Wildflower Sanctuary*. Missoula, MT: Mountain Press, 1992.

Trimble, Stephen. *The Sagebrush Ocean, a Natural History of the Great Basin*. Tenth Anniversary Edition. Reno, NV: University of Nevada Press, 1999.

### *Colorado Plateau*

Elmore, Francis H. *Shrubs and Trees of the Southwest Uplands*. Southwest Parks and Monuments Association Popular Series No. 19. Globe, AZ: Southwest Parks and Monuments Association, 1976.

Fagan, Damian. *Canyon Country Wildflowers: Including Arches and Canyonlands National Parks*. Guilford, CT: Globe Pequot Press, 1998.

Shaw, Richard J. *Utah Wildflowers: A Field Guide to the Northern and Central Mountains and Valleys*. Logan, UT: Utah State University Press, 1995.

Taylor, Ronald J. *Desert Wildflowers of North America*. Missoula, MT: Mountain Press, 1998.

Ulrich, Larry, and Susan Lamb. *Wildflowers of the Plateau and Canyon Country*. Treasure Chest Books: 1997.

Williams, David B. *Naturalist's Guide to Canyon Country*. Guilford, CT: Globe Pequot Press, 2000.

### *Columbia Plateau*

Fagan, Damian. *Pacific Northwest Wildflowers: A Guide to Common Wildflowers of Washington, Oregon, Northern California, Western Idaho, Southeast Alaska, and British Columbia*. Guilford, CT: Falcon Press, 2006.

Nisbet, Jack. *Singing Grass, Burning Sage: Discovering Washington's Shrub-Steppe*. Portland, OR: Graphic Arts Center Publishing Company, 1999.

O'Connor, Georgeanne P., and Karen J. Wieda. *Northwest Arid Lands: An Introduction to the Columbia Basin Shrub-Steppe*. Columbus, OH: Battelle Press, 2001.

Parish, Robert. *Plants of Southern Interior British Columbia and the Inland Northwest*. Vancouver, BC: Lone Pine Publishing, 1999.

Visalli, Dana, Walt Lockwood, and Derrick Ditchburn. *Northwest Dryland Wildflowers: Sagebrush-Ponderosa*. Surrey, BC: Hancock House Publishing, 2005.

Visalli, Dana, Walt Lockwood, and Derrick Ditchburn. *Northwest Mountain Wildflowers*. Surrey, BC: Hancock House Publishing, 2005.

# How to Design Native Landscapes

## Mapping and Aerial Photo Web Sites

The standard hardcopy maps to use for locating your site in the context of topography and elevation are US Geological Survey topographic maps. These are available locally in many areas. They can also be found in the map rooms of most university and some public libraries, and they can also be ordered directly from USGS: topomaps.usgs.gov/.

There are also many online sites that sell maps or charge a subscription for access to topographic and other types of maps online. These are just a few examples:

www.topozone.com
www.mapcard.com/
www.trails.com/

To put your site in geographical and topographical perspective, as well as pinpointing its location and elevation, there is no other Internet tool as good as Google Earth, which is the next best thing to a helicopter ride: earth.google.com/

## Weather and Climate Web Sites

*Western Regional Climate Center: www.wrcc.dri.edu/*

This is the repository for all the historical weather data from NOAA (National Oceanic and Atmospheric Administration) weather reporting stations throughout the western United States. It also has lots of other interesting features, as well as links to many other climate and weather sites. You just find your site on a state map or location list and click on it—you will be taken to a summary of historical weather for that site.

*Prism Group: www.prism.oregonstate.edu/*

This is cool geospatial software that has the ability to estimate climate for any location in the country. This is very handy if you are a long way from a NOAA weather reporting station or way uphill or downhill from the nearest one.

*Snotel: www.wcc.nrcs.usda.gov/snotel/*

To find out how the snow pack in the mountains of your area is doing for the current water year, visit this National Resources Conservation Service site. These people monitor snow in the mountains all across the West for the purpose of predicting water supply.

*Plant Cold Hardiness Zones: www.usna.usda.gov/Hardzone/ushzmap.html*
*Extension.usu.edu/forestry/HomeTown/Select_HardinessZoneTable.htm*

Plant cold hardiness is an important attribute determining whether a plant is likely to survive in your location. Cold hardiness is defined in terms of the coldest temperature that a plant can survive. Cold hardiness zones are based on average minimum winter temperature.

## State University Extension Web Sites

State university extension offices located in different counties throughout each state are an excellent resource for finding environmental and horticultural information specific to your area. This federal web site enables you to find the county extension office nearest you with just a couple of clicks: www.csrees.usda.gov/Extension/.

We also include web sites for university extension programs for each state in the region:

Arizona: ag.arizona.edu/extension
Colorado: www.ext.colostate.edu/
Idaho: www.extension.uidaho.edu/
Montana: extn.msu.montana.edu/
Nevada: www.unce.unr.edu/
New Mexico: extension.nmsu.edu/
Oregon: extension.oregonstate.edu/
Utah: extension.usu.edu/
Washington: ext.wsu.edu/
Wyoming: ces.uwyo.edu/

Phone numbers for local county extension offices are usually found under county government listings in local phone books.

## Resources on Soil

*State Soil Testing Labs*

Colorado State University
Fort Collins, CO 80523–1120
Phone: 970–491–5061; Fax: 970–491–2930
Email: mcschumm@lamar.colostate.edu
jimself@lamar.colostate.edu
www.colostate.edu/Depts/SoilCrop/soillab.html

New Mexico State University
Las Cruces, NM 88003
Phone: 505–646–4422; Fax: 505–646–5185
Email: wboyle@taipan.nmsu.edu
swatlab.nmsu.edu/

Oregon State University
Phone: 541–737–2187; Fax: 541–737–5725
Email: Central.Analytical.Lab@orst.edu
cropandsoil.oregonstate.edu/Services/Plntanal/CAL/

Utah State University
Phone: 435–797–2217; Fax: 435–797–2117
Email: Jkotuby@mendel.usu.edu
usual@usu.edu
www.usual.usu.edu/

University of Wyoming
Phone: 307–766–2135 (2397); Fax: 307–766–6403
Email: soiltest@uwyo.edu

*Web Resources*

Soil Texture Triangle.
soils.usda.gov/technical/manual/images/fig3–16_large.jpg

The soil texture triangle is used to determine soil texture from proportions of sand, silt, and clay. This particular graphic is part of a lengthy and rather technical online chapter on soil physical properties that is a useful resource for those who want really detailed information. The texture triangle itself is available on many different web sites.

Jar Soil Determination Method.
weather.nmsu.edu/teaching_Material/soil456/soiltexture/soiltext.htm

This site gives a more detailed description of the jar method for determining sand, silt, and clay; it also includes the texture triangle.

## Garden Design and Plant Selection

There are many perspectives on garden design for dry places—ours is only one. And there are many more native plants to choose from than the plants featured in this book. The books below, some old and some hot off the press, provide a gateway into the wider world of native plant gardening and plant selection in the American West.

### *Books*

Bornstein, Carol, David Fross, and Bart O'Brien. *California Native Plants for the Garden*. Los Olivos, CA: Cachuma Press, 2005.

Busco, Janice, and Nancy Morin. *Native Plants for High-elevation Western Gardens*. Golden, CO: Fulcrum Books, 2003.

Calhoun, Scott. *Yard Full of Sun: The Story of a Gardener's Obsession that Got a Little Out of Hand*. Tucson, AZ: Rio Nuevo Publishers, 2005.

Edwards, Betty. *The New Drawing on the Right Side of the Brain*. 2nd rev. ed. New York: Jeremy P. Tarcher / Putnam, 1999.

Keator, Glenn, and Alrie Middlebrook. *Designing California Native Gardens: The Plant Community Approach to Artful, Ecological Gardens*. Berkeley, CA: University of California Press, 2007.

Kruckeberg, Arthur. *Gardening with Native Plants of the Pacific Northwest*. 2nd ed. Seattle, WA: University of Washington Press, 1997.

Lowry, Judith L. *Gardening with a Wild Heart: Restoring California's Native Landscapes at Home*. Berkeley, CA: University of California Press, 1999.

Mee, Wendy, Jared Barnes, Roger Kjelgren, Richard Sutton, Teresa Cerny, and Craig Johnson. *Water Wise: Native Plants for Intermountain Landscapes*. Logan, UT: Utah State University Press, 2003.

Miller, George O. *Landscaping with Native Plants of Southern California*. St. Paul, MN: Voyageur Press, 2008.

Nold, Robert. *High and Dry: Gardening with Cold-hardy Dryland Plants*. Portland, OR: Timber Press, 2008.

Ogden, Scott, and Lauren Springer Ogden. *Plant-driven Design: Creating Gardens that Honor Plants, Place, and Spirit*. Portland, OR: Timber Press, 2008.

Phillips, Judith. *Natural by Design: Beauty and Balance in Southwest Gardens*. Santa Fe, NM: Museum of New Mexico Press, 1995.

Phillips, Judith. *Southwestern Landscaping with Native Plants*. Santa Fe, NM: Museum of New Mexico Press, 1987.

Robson, Kathleen, Alice Richter, and Marianne Filbert. *Encyclopedia of Northwest Native Plants for Gardens and Landscapes*. Portland, OR: Timber Press, 2008.

Smith, M. Nevin. *Native Treasures: Gardening with the Plants of California*. Berkeley, CA: University of California Press, 2006.

Springer, Lauren. *The Undaunted Garden: Planting for Weather-resilient Beauty*. Golden, CO: Fulcrum Publishing, 2000.

Tallamy, Douglas. *Bringing Nature Home: How Native Plants Sustain Wildlife in Our Gardens*. Portland, OR: Timber Press, 2007.

Tatroe, Marcia, and David Mann. *Cutting Edge Gardening in the Intermountain West*. Boulder, CO: Johnson Books, 2007.

### *Online Publications*

DeBolt, Ann, Roger Rosentreter, and Valerie Geertson, eds. *Landscaping with Native Plants of the Intermountain Region*. Technical Reference 1730–3. Boise: USDI Bureau of Land Management, Idaho State Office, 2003. www.idahonativeplants.org/guides/LandscapingGuide.aspx

Nordstrom, Sue. *Creating Landscapes for Wildlife . . . A Guide for Backyards in Utah*. Utah Division of Wildlife Resources. wildlife.utah.gov/publications/pdf/landscapingforwildlife.pdf

*Web Sites*

Ladybird Johnson Wildflower Center: www.wildflower.org/
Native Gardening and Invasive Plants Guide: enature.com/native_invasive/
Plant Native: www.plantnative.org/index.htm

# How to Water Native Landscapes

## Books and Other References

Dunnett, Nigel, and Andy Clayden. *Rain Gardens: Managing Water Sustainably in the Garden and Designed Landscape*. Portland, OR: Timber Press, 2007.

Kinkade-Levario, Heather. *Design for Water: Rainwater Harvesting, Stormwater Catchment, and Alternate Water Reuse*. Gabriola Island, BC: New Society Publishers, 2007.

Ludwig, Art. *Create an Oasis with Greywater: Choosing, Building and Using Greywater Systems, Includes Branched Drains*. 5th ed. Santa Barbara, CA: Oasis Design, 2006.

Ortho Books. *All About Sprinklers and Drip Systems*. 2nd ed. Des Moines, IA: Meredith Books. 2006.

Spence, L. E. "Root studies of important range plants of the Boise River watershed." *Journal of Forestry* 35 (1937): 747–754 (source for root drawings of native plants).

## Web Sites

Drip Irrigation: www.irrigationtutorials.com/
Graywater: www.graywater.net/
Roots: www.soilandhealth.org/01aglibrary/010137veg.roots/010137toc.html
(This is an online version of: Weaver, John E., and William E. Bruner. *Root Development of Vegetable Crops*. New York, NY: McGraw-Hill, 1927 [source of tomato root drawing].)
Water Harvesting: www.ci.tucson.az.us/water/harvesting.htm
www.harvesth2o.com/

# How to Install and Maintain Native Landscapes

## Hardscape Construction

Jeswald, Peter. *How to Build Paths, Steps, and Footbridges: The Fundamentals of Planning, Designing, and Constructing Creative Walkways in Your Home Landscape*. North Adams, MA: Storey Publishing, 2005.

Reed, David. *The Art and Craft of Stonescaping: Setting and Stacking Stone*. New York, NY: Lark Books, 2007.

Resin surfaces for paths: www.americantrails.org/resources/accessible/stabilizerstudy.html

Sagui, Pat. *Landscaping with Stone*. Upper Saddle River, NJ: Creative Homeowner, 2005.

Thompson, J. William, and Kim Sorvig. *Sustainable Landscape Construction: A Guide to Green Building Outdoors*. 2nd ed. Washington, DC: Island Press, 2007.

Trandem, Bryan, and Jerri Ferris, eds. *The Complete Guide to Creative Landscapes; Designing, Building, and Decorating Your Outdoor Home*. Minnetonka, MN: Creative Publishing International, 2000.

White, Hazel. *Hillside Landscaping*. 2nd ed. Menlo Park, CA: Sunset Books, 2007.

## Weed Control Resources

Bradley, Fern M., and Barbara W. Ellis. *Rodale's All-new Encyclopedia of Organic Gardening: The Indispensable Resource for Every Gardener*. Emmaus, PA: Rodale Books, 1993.

Corn gluten meal: www.hort.iastate.edu/gluten/ (This university web page gives excellent information on the use of corn gluten meal as a pre-emergent herbicide and provides a list of suppliers.)

Ellis, Barbara W., and Fern M. Bradley. *The Organic Gardener's Handbook of Natural Insect and Disease Control: A Complete Problem-solving Guide to Keeping Your Garden and Yard Healthy Without Chemicals*. Rev. ed. Emmaus, PA: Rodale Books, 1996.

Soil solarization: These web sites give more details on soil solarization for weed and pest control:

solar.uckac.edu/new_page1.htm

vric.ucdavis.edu/veginfo/topics/soils/soilsolarization.pdf

Whitson, Thomas, ed. *Weeds of the West*. 9th ed. Laramie, WY: Western Society of Weed Science, 2006. (The best weed identification resource for our region. Many natives meet the definition of "weed" in this book, so be sure to read the fine print.)

## Native Plant Suppliers—Internet Directories

In the quest for plant suppliers, be sure to check out the native plant society web sites listed earlier, as they often include local and state directories of native plant suppliers.

Intermountain Native Plant Growers Association: www.utahschoice.org/
Ladybird Johnson Wildflower Center: www.wildflower.org/suppliers/
Plant Native: www.plantnative.org/national_nursery_dir_main.htm

Another good source of information on plant suppliers can found on some water conservancy district web sites, which may also contain a lot of other useful information on waterwise gardening and landscaping. They may also have demonstration gardens. The water conservancy sites listed below are just some of the better examples:

Central Utah Water Conservancy District: www.centralutahgardens.org/
Jordan Valley Water Conservancy District: www.slowtheflow.org/
Weber Basin Water Conservancy District: www.weberbasin.com/wc_demo.php

## Maintenance

Lang, Susan. *Pruning*. 4th ed. Menlo Park, CA: Sunset Books, 1998.

*Aspen leaves*

# About the Authors

Dr. Susan Meyer works for the US Forest Service as a research ecologist at the Shrub Sciences Laboratory in Provo, Utah. Her specialty is restoration ecology, including seed and seedling establishment ecology and seed propagation of native plants. Past Chair of the Utah Native Plant Society State Board of Directors and one of the founding members of the Intermountain Native Plant Growers Association, she has been studying and photographing Intermountain native plants for forty years. She lives and gardens on a one-acre property in the foothills of southern Utah Valley.

Dr. Roger Kjelgren is Professor of Landscape Horticulture in the Department of Plants, Soils, and Climate at Utah State University in Logan, Utah. He is the Director of the Center for Water Efficient Landscaping, which has been a sponsor of the Intermountain Native Plant Growers Association and the Water Wise Plant Tag Program. Every other year since 1999, he has hosted a Native Plant Symposium at Utah State University, bringing interested people together to confer about the use of native plants in landscaping. He was a contributing author and guiding force for *Water Wise: Native Plants for Intermountain Landscapes*.

Darrel Morrison is a nationally known landscape architect who specializes in landscape design that is in tune with the surrounding natural environment. He was the lead landscape architect for the Ladybird Johnson National Wildflower Center in Austin, Texas. He is a Professor and Dean Emeritus of the University of Georgia School of Environmental Design and has also taught at Rutgers University (New Jersey), Utah State University (Utah), Conway School of Landscape Design (Massachusetts), and Columbia University (New York). He currently has projects in Montana, Utah, Wisconsin, California, and New York.

William Varga is Extension Horticulturist at Utah State University, where he has statewide responsibilities for outreach in low-water landscaping design. In addition to teaching around the state of Utah, he is involved with developing Utah State University Botanical Center and with other efforts to propagate native plants that have edible, medicinal, ornamental or other potential use for gardens and landscapes. He assists county agents and nursery professionals with information on plant materials for intermountain ecosystems and on the care and culture of these plants. Mr. Varga is also the author of numerous popular articles, research reports, and fact sheets on native intermountain plants.

Bettina Schultz is a stained glass artist whose work has been enriched by her experience with drawing, painting, printmaking, weaving, music, and calligraphy. She is a member of the Utah Native Plant Society and the Intermountain Native Plant Growers Association. As a founding mother of the Utah Heritage Garden Program of the Utah Native Plant Society, she has designed over a dozen native plant demonstration gardens.

*Butterfly milkweed*

# Supporters

We are grateful to the following organizations for their generous support of this book and of gardening with native plants: Utah State University Center for Water Efficient Landscaping; Utah Native Plant Society Utah Valley Chapter; Intermountain Native Plant Growers Association; Central Utah Water Conservancy District; Jordan Valley Water Conservancy District; Weber Basin Water Conservancy District; and the office of the Honorable Senator Robert Bennett.

WEBER
BASIN
WATER
CONSERVANCY
DISTRICT